SpringerBriefs in Petroleum Geoscience & Engineering

Series Editors

Jebraeel Gholinezhad, School of Engineering, University of Portsmouth, Portsmouth, UK

Mark Bentley, AGR TRACS International Ltd, Aberdeen, UK

Lateef Akanji, Petroleum Engineering, University of Aberdeen, Aberdeen, UK

Khalik Mohamad Sabil, School of Energy, Geoscience, Infrastructure and Society, Heriot-Watt University, Edinburgh, UK

Susan Agar, Oil & Energy, Aramco Research Center, Houston, USA

Kenichi Soga, Department of Civil and Environmental Engineering, University of California, Berkeley, USA

A. A. Sulaimon, Department of Petroleum Engineering, Universiti Teknologi PETRONAS, Seri Iskandar, Malaysia

The SpringerBriefs series in Petroleum Geoscience & Engineering promotes and expedites the dissemination of substantive new research results, state-of-the-art subject reviews and tutorial overviews in the field of petroleum exploration, petroleum engineering and production technology. The subject focus is on upstream exploration and production, subsurface geoscience and engineering. These concise summaries (50-125 pages) will include cutting-edge research, analytical methods, advanced modelling techniques and practical applications. Coverage will extend to all theoretical and applied aspects of the field, including traditional drilling, shale-gas fracking, deepwater sedimentology, seismic exploration, pore-flow modelling and petroleum economics. Topics include but are not limited to:

- Petroleum Geology & Geophysics
- Exploration: Conventional and Unconventional
- Seismic Interpretation
- Formation Evaluation (well logging)
- Drilling and Completion
- Hydraulic Fracturing
- Geomechanics
- Reservoir Simulation and Modelling
- Flow in Porous Media: from nano- to field-scale
- Reservoir Engineering
- Production Engineering
- Well Engineering; Design, Decommissioning and Abandonment
- Petroleum Systems; Instrumentation and Control
- Flow Assurance, Mineral Scale & Hydrates
- Reservoir and Well Intervention
- Reservoir Stimulation
- Oilfield Chemistry
- Risk and Uncertainty
- Petroleum Economics and Energy Policy

Contributions to the series can be made by submitting a proposal to the responsible Springer contact, Anthony Doyle at anthony.doyle@springer.com.

More information about this series at https://link.springer.com/bookseries/15391

Bhajan Lal · Cornelius Borecho Bavoh ·
Titus Ntow Ofei

Hydrate Control in Drilling Mud

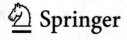 Springer

Bhajan Lal ⓘ
Department of Chemical Engineering
Universiti Teknologi PETRONAS
Seri Iskandar, Perak, Malaysia

Cornelius Borecho Bavoh
Department of Chemical Engineering
Universiti Teknologi PETRONAS
Seri Iskandar, Perak, Malaysia

Titus Ntow Ofei
Department of Geoscience and Petroleum
Norwegian University of Science
and Technology
Trondheim, Norway

ISSN 2509-3126   ISSN 2509-3134 (electronic)
SpringerBriefs in Petroleum Geoscience & Engineering
ISBN 978-3-030-94129-1   ISBN 978-3-030-94130-7 (eBook)
https://doi.org/10.1007/978-3-030-94130-7

This Springer imprint is published by the registered company Springer Nature Switzerland AG
The registered company address is: Gewerbestrasse 11, 6330 Cham, Switzerland

Preface

Drilling mud design is very crucial in all drilling operations. Drilling muds are normally adjusted to suit any drilling operation depending on the type of rock formation and the stage of the drilling operation. Different than conventional oil and gas wells where temperatures are high, gas hydrate wells have low temperatures and thus require different drilling mud properties for successful drilling. Gas hydrate wells or hydrate sediments are future reservoirs that are believed will produce clean natural gas that will replace the current fossil fuels. These methane hydrate sediments or reservoirs introduces the need for different drilling fluid systems due to their abnormal conditions (such as low temperatures conditions). Drilled cuttings from hydrate sediments are complex due to the presence of hydrates in the in-situ hydrate rock. The dissociated hydrate during drilling may reform in the wellbore or in other parts of the drilling assembly which can lead to serious drilling geo-hazard challenges. Thus, drilling mud systems used in drilling hydrate sediments must provide rheological characteristics that are capable of accommodating the presence of hydrate without blockage.

On the other hand, in offshore drilling, the flow of the drilling muds at relatively low in risers through seabed water depths provides suitable thermodynamic conditions suitable for hydrate formation in the event of a kick. This can cause serious wellbore safety and control problems during the containment of the kick. Hydrate formation incidents during offshore drilling are rarely reported in the open literature, partly because they are not recognized. There are lots of studies on designing effective drilling mud systems to enhance hydrate management in hydrate sediments drilling operations and deep offshore operations.

Hydrate management has now become a part of the drilling operation, and for that matter, relevant knowledge and guidelines of drilling mud design for hydrate management in drilling-related operation would help establish a strong foundation for hydrate-related drilling operations. In this book, we provide such information and guidelines to provide pathways and strategies for mud engineers and drilling students in the future drilling industry. The data on the effect of drilling mud additives on hydrate formation thermodynamics and kinetics is discussed to aid proper additives selection and blending for optimum performance. Almost all data on drilling mud formulation for hydrate management has been summarized with their rheo-logical properties discussed. The hydrate formation thermodynamics and kinetics data in drilling muds have been presented and discussed with guidelines to specially formulate efficient fluids for hydrate drilling activities. Practical field operations of hydrate-related drilling are discussed with insights into future drilling operations. This book is useful to mud engineers, students and industries who wish to be drilling mud authorities in the twenty-first-century energy production industry. It is worth noting that the authors strongly believe that there might be some errors in this book which could be a point of revealing truth scientifically since we know and understand in parts.

Seri Iskandar, Malaysia   Bhajan Lal
Seri Iskandar, Malaysia   Cornelius Borecho Bavoh
Trondheim, Norway   Titus Ntow Ofei

Contents

Abbreviations

3D	Three Dimensional
API	American Petroleum Institute
AV	Apparent Viscosity (mPa s)
BM	Base Mud
BOP	Blowout Preventer
Br	Bromide
BSR	Bottom Simulating Reflector
C_3H_6	Cyclopentane
CD	Carbon Dots
Cl	Chloride
CMC	Carboxymethyl Cellulose
CNT	Classical Nucleation Theory
DI	Deionized Water
DKIM	Dissociation Kinetic Inhibitive Mud
DS	Degree of Substitution
DSC	Differential Scanning Calorimetry
ECD	Equivalent Circulating Density (g/cm^3)
EOS	Equation of State
FKIM	Formation Kinetics Inhibitive Muds
FL	Fluid Loss (ml)
GA	Genetic Algorithm
Gel	Gel Strength (Pa)
GG	Guar Gum
GHSZ	Gas Hydrate Stability Zone
H_2S	Hydrogen Sulfide

H_d	Enthalpy of Dissociation (KJ/mol)
HEN	Heterogeneous
H-I-V	Hydrate–Ice–Vapor
$H-L_w-V$	Hydrate–Liquid–Vapor
HON	Homogeneous
HPHT	High Pressure High Temperature
ILs	Ionic Liquids
$I-L_w-H-V$	Ice–Liquid–Hydrate–Vapor
ind	Induction Time
KHI	Kinetic Hydrate Inhibitor
LCM	Lost Circulation Material
LDHI	Low Dosage Hydrate Inhibitors
LVER	Linear Viscoelastic Range
MD	Molecular Dynamics
MEG	Monoethylene Glycol
MeOH	Methanol
$MgCl_2$	Magnesium Chloride
MS	Modified Starch
NA Not	Applicable
$NaBr_2$	Sodium Bromide
Na-MMT	Bentonite
NGHP	National Gas Hydrate Program
NHF	No Hydrate Formed
OBM	Oil-Based Mud
PAC	Polyanionic Cellulose
PAM	Polyacrylamide
PEG	Polyethylene Glycol
PHPA	Partially Hydrolyzed Polyacrylamide
PV	Plastic Viscosity (m Pa s)
PVCap	Polyvinylcaprolactam
PVP	Polyvinylpyrrolidone
RIE	Relative Inhibition Efficiency
SD	Standard Deviation
SG	Specific Gravity
sH	Structure Three
sI	Structure One
sII	Structure Two
sol	Stable
STP	Standard Temperature and Pressure
TA	Thermal Analysis
T-cycle	Temperature Cycle
TEG/EG	Triethylene Glycol/Ethylene Glycol
THF	Tetrahydrofuran

THI	Thermodynamic Hydrate Inhibitor
TIM	Thermodynamic Inhibitive Mud
USA	United States of America
WBM	Water-Based Mud
XGUM/XG	Xanthan Gum
YP	Yield Point (Pa)

Symbols

G''	Loss Mudulus (Pa)
G'	Storage Modulus (Pa)
I	Ionic Strength
K	Consistency Index (Pa s)
M	Concnetration (Wt.%)
MW	Molecular Weight ($gmol^{-1}$)
P	Reactor Pressure (MPa)
R	Gas constant ($J\,K^{-1}\,mol^{-1}$)
$\tan\delta$	Damping Factor
t_i	Induction Time (min/h)
V	Reactor Volume (ml)
z	Compressibility Factor

Greek Letters

Δn_H	Moles of Hydrate Consumed/Uptake (mol)
ΔG	Gibbs Free Energy (kJ/mol)
η^*	Complex Viscosity (Pa s)
n	Flow Index
η	Viscosity (Pa s)
ρ	Density (g/cm^{-3})
π	Pii
γ	Shear Rate (s^{-1})
τ	Shear Stress (Pa)
σ	Sigma
$\bar{T}$	Average Depression Temperature (K/°C)

Chapter 1
Introduction

This chapter introduces the main aspects of this book by providing relevant and summarized background on the mitigation of hydrate formation in drilling mud for both deepwater drilling and hydrate sediment drilling. The need for novel and effective drilling mud systems for hydrate sediment drilling operations is also highlighted. Furthermore, the details of hydrate drilling mud systems are presented and discussed in this chapter to provide a clear overview and research progress on hydrae drilling mud systems.

1.1 Hydrates in Drilling Mud

Drilling muds are fluids employed for the drilling of any subterranean wells for water, hydrocarbon, and/or minerals production [1, 2]. From the basic knowledge of pressure in fluids, the density manipulation of the muds provides the capability to primarily control the well that is being drilled. Usually, the mud is pumped down the well via the drill string, through to the bit nozzles then up the wellbore annulus to the surface for the separation of the formation drill cuttings [3]. Most drilling mud contains water as the continuous phase or main component. Thus, when such mud systems are exposed to hydrocarbon influx through a kick and circulated through a low-temperature condition, there is a high risk of hydrate formation in the mud. This could lead to severe drilling challenges and termination of the entire drilling operation [4, 5]. In such cases, the drilling mud must be designed to manage hydrate formation risks when used in drilling hydrate sediments or deepwater drilling environments with a possible threat of hydrate formation.

© The Author(s), under exclusive license to Springer Nature Switzerland AG 2022
B. Lal et al., *Hydrate Control in Drilling Mud*,
SpringerBriefs in Petroleum Geoscience & Engineering,
https://doi.org/10.1007/978-3-030-94130-7_1

1.2 Hydrates in Offshore Drilling Mud Systems

An easier way to discuss hydrate formation in offshore drilling mud systems is briefly considered some instances of hydrate formation in drilling mud while drilling oil and gas wells. A drilling operation above 2133.6 m by Shell Oil Co.'s in the Gulf of Mexico was exposed to seabed temperatures close to 273.15 K. Considering the operating conditions with about 27.58 MPa hydrostatic pressure, hydrate easily occurred in the drilling mud systems blocking the mud circulation [6]. Similarly, Barker and Gomez [7] claimed hydrate plugged subsea equipment, causing considerable difficulties in subsequent operations. The formation of the hydrate in offshore drilling mud systems might cause several challenges such as;

- Ram cavity blockage
- Annular blockage between the BOP and the drillstring.
- Blockage of chocklines and kill lines.
- BOP blockage or plugging.

Therefore, to avoid the above-mentioned challenges, drilling mud systems for drilling hydrate-prone environments must be able to prevent hydrate formation. Therefore, the drilling mud systems contain other additives, the hydrate inhibitive additives must be compatible with most conventional drilling mud additives used for drilling.

1.3 Hydrate Sediments Drilling Muds—The New Paradigm

The distribution of natural methane hydrates around the earth, especially in marine sediments and permafrost environments allows the exploitation of methane for future energy [8]. The amount of energy trapped as hydrates doubles that of fossil fuels [5]. Thus, producing this trapped methane requires a new kind of drilling technology. Hence, the drilling mud design would be shifting from high-temperature wells to low-temperature wells applications. This further calls for mud properties that are stable at low temperatures and can manage hydrate sediment rock hydrate-related challenges. Therefore, the new type of drilling mud is a paradigm shift from conventional drilling mud systems. When exploring and producing methane hydrates, much attention must be given to its drilling techniques, with a special focus on the drilling mud technology. This is because drilling hydrate sediments might cause disturbances on the situ hydrate rock. Which could destabilize or melt the methane hydrates in the sediments. The melting of the hydrates during drilling can lead to series of severe drilling geohazards and wellbore instability [9, 10]. When there is a wellbore instability in an uncased wellbore scenario the melted hydrates would release gas into the seafloor via the drilling pipe. Also, the gas release in the drilling mud might plug the drilling pipe and stop mud circulation due to hydrate reformation in the drill pipe.

In severe instances, more gas release in the wellbore might lead to a blowout or platform collapse due to seawater gasification. In cased wellbore, there could be a casing collapse due to high pressure associated with the dissociated hydrates [11, 12]. To efficiently prevent these challenges, a careful choice and preparation of drilling mud is the best way in current practices. Hence, a new paradigm drilling mud system design is suitable to manage hydrate drilling geohazards for safe operations in drilling hydrate zones. To effectively manage hydrate while drilling, the desired drilling mud system must have a suitable relative density (ρ) variation window depending on the depth or location of the hydrate zone being drilled. This would also ensure the propagation pressure needed to maintain the wellbore stability by preventing hydrate dissociation via depressurization. According to Jiang et al. [13], relative mud density in the range of 1.05–1.2 is suitable for drilling mud sediments. In addition to the suitable density range, the hydrate drilling mud system must be rheologically stable at low temperatures, also, the temperature of the mud must be thermally stable to avoid hydrate dissociation in the in-situ rock. Most importantly, the drilling mud should efficiently manage hydrate reformation and dissociation in the wellbore drilling assembly, and blowout preventer. Also, the mud must possess low filtration and should be good enough to lubricate the drilling assembly. Last but not least, hydrate drilling mud should be designed to prevent calcium and magnesium-ion pollution [13].

1.4 Summary of Hydrate Studies in Drilling Mud

This section provides a general overview of all the studied drilling mud systems on hydrate sediment drilling operation and hydrate mitigation in offshore drilling activities. Details of the results and findings for each study are discussed in Chaps. 4–6. By far about 37 articles [6, 13, 15–19, 21, 23–49] are reviewed and presented in Table 1.1. Most of the authors either study the phase behavior [6, 14], kinetics [15–17], and dissociation of hydrates in the presence of drilling mud with different additives [18, 19]. Some empirical modeling and simulation analysis has also been conducted on some systems using the conventional hydrate phase behavior electrostatic and alcohols prediction models. Software such as HWHYD model, CSMHYD simulation, MD simulation, and statistical models has been used by authors to study the effect of drilling mud in the presence of hydrates [6, 20, 21]. Details of the hydrate models can be found in Sloan Book[22] and our previous book, thus, this current book won't focus much on the models, but on the experimental data and findings in literature. Typical drilling mud hydrate-based studies are conducted in natural gas systems with high methane content (>85%). Other gas components in the natural gas systems include propane, ethane, butane, pentane, hexane, carbon dioxide, and nitrogen. Other studies use pure methane, THF, and ice. Considering the guest molecules being studied by authors, structure I and structure II hydrate systems are the mainly focused systems.

Some of the natural gas systems tested are real field gas samples for the Gulf of Mexico and other locations. Water-base mud systems are mostly used for hydrate studies with few on oil-based mud systems and spot fluids. Some also test the effect of drilling mud additives on gas hydrate formation using deionized and freshwater systems. The base fluid systems normally consist of polymers, salts, and water at different fractions prepared to specific densities and pH.

Basic gas hydrate autoclave cell is commonly used to test the hydrate formation/dissociation kinetics in drilling muds. The same apparatus is employed for the hydrate phase behavior test by employing the pressure search method. The use of DSc and simulation software has also been used for phase behavior studies. Other apparatus such as the Fann viscometer is used to study the rheological properties of the mud. This is normally performed to study the effect of the hydrate additive on the drilling mud properties. However, using modern rheological equipment such as rheometers could be more accurate and efficient. In detail, Chap. 3 discusses the methods and procedures used for testing the hydrate and rheological properties of hydrate drilling mud systems. Gas hydrate additives such as salts, glycols, polymers, and ionic liquids are used to either prevent hydrate formation or dissociation up to 40 wt%.

1.5   Hydrate Behavior in Drilling Mud Additives

The basic knowledge of hydrate behavior in drilling mud is the key to efficiently manage their possible threat during drilling and mud formulation. Generally, hydrates formation can be viewed from two chemical engineering perspectives as kinetics and thermodynamics. Thus, one must be well vested in the kinetics and thermodynamic of gas hydrate formation to fully understand its management in drilling mud. The thermodynamics or phase behavior of hydrate is dependent on the guest or gas molecule present and the size and shape of the molecules [22]. Based on this, several types of hydrate could be formed, since hydrates in drilling mud mainly deal with methane and natural gas, structure I and II are most likely to occur. There could be a possibility of forming structure H hydrate but is very rare and only possible if there is a kick in the light hydrocarbon formation which has structure H hydrate forming molecules [22]. The thermodynamics of hydrate provided the basic information on when (pressure and temperate) hydrate could form in any system. An afore-knowledge on this could help prevent hydrate formation, especially in deepsea drilling operations. The phase behavior of hydrates in drilling mud systems is discussed in detail in Chap. 4.

On the other hand, the kinetics of hydrate formation mainly deals with the hydrate formation path ranging from nucleation to crystal growth. The kinetics study measures the time taken for hydrates to form (known as induction or nucleation time), the rate for hydrate formation, and the amount of moles of gas consumed in hydrates. Kinetic hydrate inhibitors mostly act to delay the hydrate formation time and rate of formation, as well as reducing the amount of gas consumed into hydrate [9]. Anti-agglomerate hydrate inhibitors are different types of kinetic hydrate inhibitors the prevent hydrate crystals from aggregating to plug a system. The kinetics of hydrate dissociation is also a very important behavior to evaluate. It provides information on the possible release of gas into the wellbore when the mud comes to contact with the hydrates. The time to release a certain amount of gas can be determined and prevented when formulating the drilling mud system. Details and fundamentals on hydrate formation thermodynamics and kinetics are discussed in Sloan's Book [22].

1.6   Significance of Managing Hydrate in Drilling

1.6.1   Safe Offshore Drilling Operation

Safe operation of drilling activities is the goal of any drilling activity. In deepsea drilling operations, hydrate management could save time, cost, and lives. For instance, in 2010, poor gas hydrate management was the root cause of the oil/gas well blowout of the Macondo well in the Gulf of Mexico, which lead to the death of 11 crew members [50]. Thus, to ensure smooth and hydrate-free drilling, efficient hydrate management techniques and drilling mud design are required.

1.6.2   Hydrate Sediment Drilling for Energy and CO_2 Sequestration

The promising energy potentials in hydrate sediments can only be possible if the sediments are well drilling without challenges [8]. The key to effectively drilling hydrate sediments lies in the drilling mud system's ability to effectively manage the dissociation of the in-situ hydrate and as well prevent the reformation of hydrate in the wellbore owing to the presence of the dissociated gas in the mud system. By so doing, the mud would be able to play its perfect role and carry drilled cuttings to the surface with good wellbore stability and cleaning effect. We believe that the successful drilling of hydrate sediments would set a high and promising advantage for producing energy from hydrate sediments around the globe.

1.7   Structure of the Book

This book is structured to provide an easy flow and understanding of the subject area. The chapters are well organized to discuss the objectives of the book, separately discussing standalone sections that are relevant to readers. Readers may choose to read specific chapters or decide to flow with the entire book. This current chapter, which is the first, introduces the book and provides background knowledge on gas hydrate drilling geohazards and wellbore instability challenges associated with hydrate sediment drilling. Deepwater drilling challenges related to hydrate and the properties of suitable drilling mud systems to manage hydrate formation while drilling is explained. The second chapter discusses the fundamentals of hydrate formation and hydrate sediments. The types of drilling muds and their formulation additives regarding hydrate management are also discussed to well understand the proceeding chapters. The methods and procedures for testing hydrate management drilling mud properties are described in Chap. 3. The chapter discusses include rheology testing methods and hydrate formation and dissociation testing procedures. Chapter 4 is dedicated to the discussion on the phase behavior results on hydrate drilling mud additives. It also presents significant phase behavior data on hydrate drilling mud. Chapter 5 discusses the kinetics results on hydrate drilling mud additives. It also presents significant kinetic data on hydrate drilling mud. Chapter 6, on the other hand, presents a discussion on the reported rheological data on hydrate drilling muds. The last chapter, which is Chap. 7, gives a brief practical field situations scenarios of gas hydrate formation in drilling mud systems while drilling.

Table 1.1 Summary of studies on hydrate management in drilling mud systems

Authors	Type of study	Gas	Type of Mud	Methods	Apparatus	Hydrate additives	P (MPa)/T (K)	Rheology study	Remarks
Zhao et al. [23]	Hydrate dissociation kinetics	CH_4	WBM	Constant cooling and heating with injection	Autoclave cell and a viscometer	PVP soybean lecithin, and EG	10/275–288.15	AV, PV, YP, Gel, and FL	Both PVP and lecithin can delay methane hydrate dissociation while drilling. However, their combination is much better and compatible with the drilling mud. Using EG at a concentration below 10 wt% as a THI instead of NaCl can also help inhibit hydrate dissociation
Srungavarapu et al. [24]	Hydrate Kinetics	CH_4	WBM	Constant cooling	Autoclave cell and a viscometer	CMC, XG	16.55/271.4–303	AV, PV, YP, YP/PV ratio, Gel, and FL	XG and CMC mitigate CH_4 hydrate formation in both static and dynamic conditions. However, the presence of a kick retards their performance
Fereidounpour and Vatani [25]	Hydrate dissociation	THF	WBM	Constant cooling and heating with injection	Hydrate cell and a viscometer	Pre-hydrated bentonite, polymer, KCL/PHPA. And Advanced polymer	0.1/273–293.15	PV, YP, gel and MW, and FL	The presence of polymers such as PHPA and PACR in drilling mud reduces hydrate dissociation.

(continued)

Table 1.1 (continued)

Authors	Type of study	Gas	Type of Mud	Methods	Apparatus	Hydrate additives	P (MPa)/T (K)	Rheology study	Remarks
Liyi et al. [17]	Hydration Expansion	AKASS	WBM	Constant expansion	JULABO Laortechnik GmbH 77,960 Seelbach	HT, SHR, SMT, and FCLS	20/263.15	PV, YP, gel and MW, and FL	HT provides suitable drilling mud rheological and filtration properties for low-temperature drilling activities. In addition, potassium ion prevents borehole collapse when drilling low-temperature wells
Østergaard et al. [21]	Hydrate phase behavior	CH4 and NG1	WBM	HWHYD model	Software	HCOOK, NaCl, NaBr, KCl, $CaCl_2$, $MgCl_2$, glycerol, MeOH, KCl, EtOH, $SrCl_2$, MEG, TEG, DEG, KBr, $BaCl_2$, Na_2SO_4	3.45–31.72/275.8–278.7	NA	Thermodynamic models are capable of predicting the hydrate-free zone in the presence of salts and chemical inhibitors for drilling activities
Fereidounpour and Vatani [26]	Hydrate dissociation	THF	WBM	Constant cooling and heating with injection	Hydrate cell and a viscometer	Polyacrylates Advanced polymer	0.1/273–293	PV, YP, gel and MW, and FL	The Polyacrylate fluids reduces the heat exchange rate to prevent gas hydrate dissociation

(continued)

Table 1.1 (continued)

Authors	Type of study	Gas	Type of Mud	Methods	Apparatus	Hydrate additives	P (MPa)/T (K)	Rheology study	Remarks
Mech and Sangwai [27]	Hydrate formation / dissociation kinetics	CH$_4$	DI	Constant cooling and heating	high-pressure reactor	PEG 200 and PEG 600	5.5–7.5/263–303	NA	PEG shows a higher CH$_4$ hydrate kinetics inhibition impact. Their inhibition effect is controlled by their molecular weight
Jiang et al. [13]	Hydrate formation kinetics	CH$_4$	WBM	Constant cooling	Autoclave and viscometer	MgCl2, CaCl2, PVP K-90	18/265–288	YP, PV, gel, FL	Polyethylene glycol with a high percentage of NaCl and low addition of PVP K-90 is suitable to reduce hydrate formation with good rheological properties at low temperatures
Zhao et al. [19]	Wellbore stability	NA	NA	Permeability and compressive stress measurement	Penetrometer and pointer-type tensimeter setup	KCl, NaCl, MgCl2, Na2SO4, and CaCl2, PACLV, PHPA, Al-seal, SDJA	0.7–2.5/298.15	NA	Both silicates and Al-seal have favorable cementation and consolidation effects on the wellbore and can enhance wellbore stability in shallow formation under deep water

(continued)

Table 1.1 (continued)

Authors	Type of study	Gas	Type of Mud	Methods	Apparatus	Hydrate additives	P (MPa)/T (K)	Rheology study	Remarks
Yu et al. [28]	Hydrate dissociation	CH_4	NA	Cooling and heating	High-pressure hydrate reactor	NA	7/275	NA	Hydrate dissociation and gas flow into wellbore can be induced by the circulation of high-temperature drilling mud when drilling through hydrate-bearing sediments. The experimental results show that the rates of hydrate dissociation and gas production are greatly influenced by the temperature of drilling mud and hydrate saturation and pressure
Zhang et al. [18]	Hydrate dissociation	CH_4	WBM	Cooling and heating by injection of drilling mud	Custom design reactor	NaCl and EG	5–13/273–278	YP, PV, and FL	High drilling mud temperatures cause THIs to significantly promote hydrate dissociation with increasing concentration

(continued)

Table 1.1 (continued)

Authors	Type of study	Gas	Type of Mud	Methods	Apparatus	Hydrate additives	P (MPa)/T (K)	Rheology study	Remarks
Xiaolan et al. [29]	Hydrate phase behavior	CH_4	NA	T-cycle	Autoclave	PF-THIN, SD-102, XY-27, PAM, and LV-CMC	Not stated	NA	Polyalcohol-type chemicals were found to have significant hydrate inhibition effects due to their multiple hydroxyl groups. Sulfonated methyl tannin (PF-THIN) was found to have some promotive effect on hydrate formation
Kawamura et al. [30]	Hydrate dissociation	$90.1\&CH_4 + 9.9\%C_2H_6$	WBM	Isothermal and Isobaric methods	High-pressure vessel	XANVIS	8.5–10/270.15	Shear rate and viscosity	The dissociation rate of gas hydrates in drilling muds is almost proportional to the concentration of mud then it can be suggested that in the case of riser drilling, the dissociation rate will be slow compared with the case of no drilling mud

(continued)

Table 1.1 (continued)

Authors	Type of study	Gas	Type of Mud	Methods	Apparatus	Hydrate additives	P (MPa)/T (K)	Rheology study	Remarks
Jiang et al. [15]	Hydrate formation kinetics	CH_4	WBM	Constant cooling	High-Pressure autoclave	Polyglycol and PVP K90	13.8/273.15	PV, YP	Clay promotes hydrate formation, while modified starch and polyglycols can inhibit the formation of hydrate to a certain extent. PVP K90 has an excellent inhibition effect on hydrate formation in drilling mud
Jiang et al. [13]	Hydrate formation kinetics	CH_4	WBM	Constant cooling	Autoclave and viscometer	$MgCl_2$, $CaCl_2$, PVP K-90	18/265–288	YP, PV, gel, FL	Polyethylene glycol with a high percentage of NaCl and low addition of PVP K-90 is suitable to reduce hydrate formation and provide good rheological properties at low temperatures

(continued)

Table 1.1 (continued)

Authors	Type of study	Gas	Type of Mud	Methods	Apparatus	Hydrate additives	P (MPa)/T (K)	Rheology study	Remarks
Liu et al. [31]	Physicochemical analysis	NA	NA	NA	NA	DSO-3,0cs, KF96-2,0cs, ethyl butyrate, n-propyl propionate, n-butyl butyrate, and n-amyl butyrate	0.1/258–298	NA	low-molecular-weight dimethylsiloxane oils (DSOs) and low molecular weight fatty-acid esters (FAEs) has potential properties as drilling mud additives for drilling hydrate sediments and arctic areas
Yan et al. [32]	Hydrate phase behavior	CH$_4$	–	Constant volume method and MD simulation	High-pressure crystallizer	Montmorillonite	5.1/275.15	NA	Methane hydrate in montmorillonite forms sI hydrates with both hydrate formation occurring in the bulk and within the montmorillonite layers
Nikolaev et al. [33]	Hydrate formation kinetics	CH$_4$	WBM	Constant cooling	High pressure autoclave	PVP K90	8/274.15	PV, YP, Gel, FL, density	Composite alcohol drilling mud has a reasonable density, good low-temperature rheology, lubricity, shale hydration inhibition, and hydrate formation inhibition properties
Ning et al. [16]	hydrate formation	CH$_4$	OBM	Constant cooling	High pressure autoclave	Ethylene glycol	20/277.15	NA	OBMs are prone to hydrate formation, especially at very high water cuts (> 10 wt.%). The presence of ethylene glycol could prevent hydrate formation in OBM

(continued)

Table 1.1 (continued)

Authors	Type of study	Gas	Type of Mud	Methods	Apparatus	Hydrate additives	P (MPa)/T (K)	Rheology study	Remarks
Ning et al. [34]	Hydrate formation kinetics and phase behavior modeling	CH_4	OBM	Constant cooling and empirical modelling	Autoclave, CSMHyd, HWHYD	EG	20/277.15	NA	Thermodynamic inhibitors and solid-phase still play a decisive role in estimating the hydrate safety margin of offshore drilling muds. The existing hydrate behavior prediction models for salts and alcohols are applicable in OBM systems
Dzialowski et al. [35]	Hydrate formation kinetics	Green Canyon gas composition	WBM	Constant cooling and heating	High-pressure cell	EG	15–31/269–301	NA	Kinetic inhibitors could be used to prevent hydrate formation in WBM mud systems. However, it's better to combine them with THIs
Saikia and Mahto [36]	Hydrate formation kinetics	THF	WBM	Constant cooling	Clathrate hydrate-crystallizer	PVP, ionic liquid, 1-decyl-3-methylimidazolium tetrafluoroborate	0.1/275–293	Viscosity, share stress and shear rate	Ionic liquid (1-decyl-3-methylimidazolium tetrafluoroborate) can inhibit THF hydrate crystals comparable to PVP
Saikia and Mahto [37]	Hydrate formation kinetics	THF	WBM	Constant cooling	High-pressure cell	PVP, lipase	0.1/275	AP, PV, YP	Lipase shows good hydrate inhibition and anti-agglomerate characteristic, however, its kinetic inhibition impact is a similar hydrate inhibition property as PVP

(continued)

Table 1.1 (continued)

Authors	Type of study	Gas	Type of Mud	Methods	Apparatus	Hydrate additives	P (MPa)/T (K)	Rheology study	Remarks
Saikia and Mahto [38]	Rheological behavior	–	WBM	Statistical method	RSM and CCD	EG, DEG, PVP, PVCap, VC713	0.1/275–303	Viscosity	Drilling mid rheology is more sensitive to temperature than the concentration of hydrate inhibitors
Saikia et al. [39]	Hydrate formation kinetics	THF	WBM	Constant cooling	High-pressure cell	EG and carbon dots (CDs)	0.1/277.5	AP, PV, YP	Carbon dots could inhibit clathrate hydrate better than ethylene glycol at low concentrations. However, Carbon dots has a negligible effect on drilling mud properties
Lai and Dzialowski [40]	Hydrate phase boundary and kinetics	NG2	WBM	T-cycle and constant cooling	High-pressure cell	Subject to mud composition	3–37.8/274–300	NA	Lignosulfonate and lignite muds do not inhibit hydrate formation. NaCl, glycerine, and propylene glycol depress the hydrate equilibrium temperature and slow down the reaction rate if concentration levels are high. Caustic, gel, diesel, and calcium have minor effects on the hydrate equilibrium point and reaction rate
Hale and Dewan [41]	Hydrate phase behavior	NG3	WBM	Freezing point depression estimations	CSMHYD simulation	NaCl and Glycerol	6.7–31/288–366	AV, PV, YP, gel, FL	NaCl/glycerol-based drilling mud systems can thermodynamically prevent hydrate formation in drilling operations

(continued)

Table 1.1 (continued)

Authors	Type of study	Gas	Type of Mud	Methods	Apparatus	Hydrate additives	P (MPa)/T (K)	Rheology study	Remarks
Grigg and Lynes [42]	Hydrate phase behavior	NG4	OBM	T-cycle method	High-pressure blind cell	$CaCl_2$	2.5–31.8/274–300	PV, YP, gel	Oil as a continuous phase in an oil-based drilling mud does not prevent gas-hydrates formation, rather it reduces the gas-hydrates formation prone region. The addition of CaCl2 brine further reduces the risk of gas-hydrates formation in Oil-based drilling muds.
Kotkoskie et al. [6]	Hydrate phase behavior	NG5	WBM	T-cycle method	High-pressure blind cell	NaCl, NaBr, $CaCl_2$	2.5–28.9/277–301.8	NA	The type of salt concentration controls the prediction of hydrate formation in drilling mud. The presence of some mud additives (excluding salts) slightly promotes hydrate formation
Saikia and Mahto [37]	Hydrate formation kinetics	THF	WBM	Constant cooling	High-pressure cell	PVP, lipase	0.1/275	AP, PV, YP	Lipase shows good hydrate inhibition and anti-agglomerate characteristic, however, its kinetic inhibition impact is similar to PVP

(continued)

Table 1.1 (continued)

Authors	Type of study	Gas	Type of Mud	Methods	Apparatus	Hydrate additives	P (MPa)/T (K)	Rheology study	Remarks
Ebeltoft et al. [43]	Hydrate phase behavior	NG6	WBM	T-cycle method and empirical modelling	High-pressure cell	EG, NaBr, NaCl, KCl, ammonium Calcium nitrate, Na-formate, CaCl2	10.3–37.9/274–298	NA	NaCl is the most effective hydrate inhibitor, followed by KCl, CaCl2, NaBr, Na-formate, and calcium nitrate. Ethylene glycol is the best glycol inhibitor, Na-formate is better than NaCl but precipitates at low temperatures and high concentrations
Schofield et al. [44]	Hydrate formation kinetics	NG7	WBM	Constant cooling method	High-pressure cell	Lecithin, glycerol, PVP, ConDet, Defoamer	10.3 /280.35	PV. YP, Gel, FL	Lecithin kinetically keeps hydrates stable and alters drilling mud rheological properties
Herzhaft and Dalmazzone [45]	hydrate phase behavior and kinetics	CH$_4$ and NG8	OBM and WBM	Isothermal mode	DSC	Carbonate solids, CaCl2 and NaCl	0.1–9/234–302.9	NA	WBM or OBM hydrate-based studies could be studied using the DSc method with good agreement. Including the presence of inhibitors and solids

(continued)

Table 1.1 (continued)

Authors	Type of study	Gas	Type of Mud	Methods	Apparatus	Hydrate additives	P (MPa)/T (K)	Rheology study	Remarks
Zhao et al. [46]	hydrate formation	NG9	WBM	Constant cooling	High-pressure cell	PVP, glycerol, glycol, NaCl	10 MPa/275–289.5	PV and YP	The hydrate inhibition efficiency of PVP is not very significant. However, the NaCl synergically enhances the hydrate inhibition effect of PVP. Glycol and glycerol do not synergically prevent hydrate with PVP
Gupta et al. [47]	Hydrate phase behavior	CH_4	NA	T-cycle	High Pressure cell	XG, PAM, GG	5.5–8.6/279.6–284	NA	Water-soluble polymers slightly ($\Delta T \leq 1.05$ K) inhibits CH_4 hydrates thermodynamically depending on their molecular weight

(continued)

Table 1.1 (continued)

Authors	Type of study	Gas	Type of Mud	Methods	Apparatus	Hydrate additives	P (MPa)/T (K)	Rheology study	Remarks
Wang et al. [48]	Low-temperature rheology	NA	WBM	API standards	Viscometer	Semen Lepidii natural vegetable gum KL, PAM, and XC	0.101/263–284	Viscosity, AV, YP	The presence of KL, PAM, and XG in WBM enhanced their low-temperature response by slightly changing their rheological properties. In addition, the Herschel–Bulkley model is suitable for studying hydrate sediment drilling mudss
Wang et al. [49]	Hydrate formation kinetics	CH_4	NA	Constant cooling	Hydrate reaction simulator	MS, CMC, XG	5–12/278.15	AV, PV	XG inhibits CH_4 hydrate formation better than MS, CMC. However, their performance is challenged at a higher driving force

$NG1$ (87.29C1 + 7.55C2 + 3.09C3 + 049i-C4 + 0.79n-C4 + 0.39n-C5 + 0.4N2); Aqueous kilndried artificial sodium soil (AKASS); Deionized water (DI); Water-based mud (WBM); Low-viscosity polyanionic cellulose (PACLV); Partially hydrolyzed polyacrylamide (PHPA); Aluminium complex (A1-seal); Sodium silicate and potassium silicate; Polyamine (SDIA)
$NG2$ (87.55C1 + 7.59C2 + 3.1C3 + 0.19i-C4 + 0.79n-C4 + 0.19i-C5 + 0.19n-C6 + 0.39N2); $NG3$ (87.29C1 + 7.55C2 + 3.09C3 + 0.49i-C4 + 0.79n-C4 + 0.39i-C5 + 0.4N2);
$NG4$ (87.55C1 + 7.59C2 + 3.1C3 + 0.19i-C4 + 0.79n-C4 + 0.19i-C5 + 0.19n-C6 + 0.39N2); $NG5$ (87.2C1 + 7.6C2 + 3.1C3 + 0.5i-C4 + 0.8n-C4 + 0.2i-C5 + 0.2n-C5 + 0.4N2);
$NG6$ (87.243C1 + 7.57C2 + 3.08C3 + 0.51i-C4 + 0.792n-C4 + 0.202i-C5 + 0.2nC5 + 0.403N2).
$NG7$ (94C1 + 4C2 + 2C3); $NG8$ (94.3C1 + 2.58C2 + 0.57C3 + 0.09i-C4 + 0.08n-C4 + 0.24 CO_2 + 2.14 N_2); $NG9$ (80.42C1 + 10.35C2 + 1.82 CO_2 + 0.11 N_2); NA (Not applicable).

References

1. Caenn R, Chillingar GV (1996) Drilling fluids: state of the art. J Pet Sci Eng 14:221–230
2. Caenn R, Darley HCH, Gray GR (2017) The rheology of drilling fluids. In: Compos. prop. drill. complet. fluids. Elsevier Inc, pp 151–244
3. Caenn R, Darley HCH, Gray GR (2017) Drilling fluid components. In: Compos. prop. drill. complet. fluids. Elsevier Inc, pp 537–595
4. Kim NR, Ribeiro PR, Bonet EJ (2007) Study of hydrates in drilling operations: a review. Braz J Pet Gas 1:116–122
5. Merey Ş (2016) Drilling of gas hydrate reservoirs. J Nat Gas Sci Eng 35:1167–1179
6. Hale H, Dewan AKR (1990) Inhibition of gas hydrates in deepwater drilling. SPE Drill Eng 7:109–115
7. Barker JW, Gomez RK (1989) Formation of hydrates during deepwater drilling operations. J Pet Technol 41:297–301
8. Englezos P (1993) Clathrates hydrates. Ind Eng Chem Res 32:1251–1274
9. Bavoh CB, Lal B, Keong LK (2020) Introduction to gas hydrates. Chem Addit Gas Hydrates. https://doi.org/10.1007/978-3-030-30750-9_1
10. Bavoh CB, Ofei TN, Lal B, Sharif AM, Shahpin MHBA, Sundramoorthy JD (2019) Assessing the impact of an ionic liquid on NaCl/KCl/polymer water-based mud (WBM) for drilling gas hydrate-bearing sediments. J Mol Liq 294:111643
11. Ghajari MP, Sabkdost A (2014) Hydrate-related drilling hazards and their remedies. In: 2nd Natl. Iran. conf. gas hydrate, Semnan Univ.
12. Bavoh CB, Ofei TN, Lal B (2020) Investigating the potential cuttings transport behavior of ionic liquids in drilling mud in the presence of sII hydrates. Energy Fuels 34:2903–2915
13. Jiang G, Liu T, Ning F, Tu Y, Zhang L, Yu Y, Kuang L (2011) Polyethylene glycol drilling fluid for drilling in marine gas hydrates-bearing sediments: an experimental study. Energies 4:140–150
14. Bavoh CB, Md Yuha YB, Tay WH, Ofei TN, Lal B, Mukhtar H (2019) Experimental and modelling of the impact of quaternary ammonium salts/ionic liquid on the rheological and hydrate inhibition properties of xanthan gum water-based muds for drilling gas hydrate-bearing rocks. J Pet Sci Eng 183:106468
15. Jiang G, Ning F, Zhang L, Tu Y (2011) Effect of agents on hydrate formation and low-temperature rheology of polyalcohol drilling fluid. J Earth Sci 22:652–657
16. Ning F, Zhang L, Tu Y, Jiang G, Shi M (2010) Gas-hydrate formation, agglomeration and inhibition in oil-based drilling fluids for deep-water drilling. J Nat Gas Chem 19:234–240
17. Chen L, Wang S, Ye C (2014) Effect of gas hydrate drilling fluids using low solid phase mud system in plateau permafrost. Procedia Eng 73:318–325
18. Zhang H, Cheng Y, Shi J, Li L, Li M, Han X, Yan C (2017) Experimental study of water-based drilling fl uid disturbance on natural gas hydrate-bearing sediments. J Nat Gas Sci Eng 47:1–10
19. Zhao X, Qiu Z, Wang M, Xu J, Huang W (2019) Experimental investigation of the effect of drilling fluid on wellbore stability in shallow unconsolidated formations in deep water. J Pet Sci Eng 175:595–603
20. Yousif MH, Young DB (1993) Simple correlation to predict the hydrate point suppression in drilling fluids. In: Drill conf—proc, pp 287–294
21. Østergaard KK, Tohidi B, Danesh A, Todd AC (2000) Gas hydrates and offshore drilling. predicting the hydrate free zone. Ann N Y Acad Sci 912:411–419
22. Sloan ED, Koh CA (2008) Clathrate hydrates of natural gases, 3rd edn. CRC Press Taylor & Francis, Boca Raton; London; New york
23. Zhao X, Qiu Z, Zhao C, Xu J, Zhang Y (2019) Inhibitory effect of water-based drilling fluid on methane hydrate dissociation. Chem Eng Sci 199:113–122
24. Srungavarapu M, Patidar KK, Pathak AK, Mandal A (2018) Performance studies of water-based drilling fluid for drilling through hydrate bearing sediments. Appl Clay Sci 152:211–220
25. Fereidounpour A, Vatani A (2014) An investigation of interaction of drilling fluids with gas hydrates in drilling hydrate bearing sediments. J Nat Gas Sci Eng 20:422–427

26. Fereidounpour A, Vatani A (2015) Designing a Polyacrylate drilling fluid system to improve wellbore stability in hydrate bearing sediments. J Nat Gas Sci Eng 26:921–926
27. Mech D, Sangwai JS (2016) Effect of molecular weight of polyethylene glycol (PEG), a hydrate inhibitive water-based drilling fluid additive, on the formation and dissociation kinetics of methane hydrate. J Nat Gas Sci Eng 35:1441–1452
28. Yu L, Xu Y, Gong Z, Huang F, Zhang L, Ren S (2018) Experimental study and numerical modeling of methane hydrate dissociation and gas invasion during drilling through hydrate bearing formations. J Pet Sci Eng 168:507–520
29. Xiaolan L, Zuohui L, Yong Z, Rongqiang L (2008) Gas hydrate inhibition of drilling fluid additives. In: 7th Int. conf. gas hydrates, edinburgh, Scotland, United Kingdom, pp 1–6
30. Kawamura T, Yamamoto Y, Yoon JH, Haneda H, Ohga K, Higuchi K (2002) Dissociation behavior of methane-ethane mixed gas hydrate in drilling mud fluid. Proc Int Offshore Polar Eng Conf 12:439–442
31. Liu N, Xu H, Yang Y, Han L, Wang L, Talalay P (2016) Physicochemical properties of potential low-temperature drilling fluids for deep ice core drilling. Cold Reg Sci Technol 129:45–50
32. Yan K, Li X, Chen Z, Zhang Y, Xu C, Xia Z (2019) Methane hydrate formation and dissociation behaviors in montmorillonite. Chin J Chem Eng 27:1212–1218
33. Nikolaev NI, Liu T, Wang Z, Jiang G, Sun J, Zheng M, Wang Y (2014) The experimental study on a new type low temperature water-based composite alcohol drilling fluid. Procedia Eng 73:276–282
34. Ning F, Zhang L, Jiang G, Tu Y, Wu X, Yu Y (2011) Comparison and application of different empirical correlations for estimating the hydrate safety margin of oil-based drilling fluids containing ethylene glycol. J Nat Gas Chem 20:25–33
35. Dzialowski A, Patel A, Nordbo MLLCK, Hydro N The development of kinetic inhibitors to suppress gas hydrates in extreme drilling conditions. In: Offshore mediterr. conf. exhib. Ravenna, Italy, pp 1–15
36. Saikia T, Mahto V (2016) Evaluation of 1-decyl-3-methylimidazolium tetrafluoroborate as clathrate hydrate crystal inhibitor in drilling fluid. J Nat Gas Sci Eng 36:906–915
37. Saikia T, Mahto V (2016) Experimental investigations of clathrate hydrate inhibition in water based drilling fluid using green inhibitor. J Pet Sci Eng 147:647–653
38. Saikia T, Mahto V (2018) Experimental investigations and optimizations of rheological behavior of drilling fluids using RSM and CCD for gas hydrate-bearing formation. Arab J Sci Eng 43:1–14
39. Saikia T, Mahto V, Kumar A (2017) Quantum dots: a new approach in thermodynamic inhibitor for the drilling of gas hydrate bearing formation. J Ind Eng Chem 52:89–98
40. Lai DT, Dzialowski AK (1989) Investigation of natural gas hydrates in various drilling fluids. In: SPE/IADC drill conf New Orleans, Louisiana 18637, pp 181–194
41. Hale AH, Dewan AKR (1990) Inhibition of gas hydrates in deepwater drilling. SPE Drill Eng 5(109–115):18638
42. Grigg RB, Lynes GL (1992) Oil-based drilling mud as a gas-hydrates inhibitor. SPE Drill Eng 7:32–38
43. Ebeltoft H, Yousif M, Westport I (2001) Hydrate control during deepwater drilling: overview and new drilling-fluids formulations, pp 5–8
44. Schofield TR, Judzis A, Yousif M (1997) Stabilization of in-situ hydrates enhances drilling performance and rig safety. In: SPE Annu. tech. conf. exhib, pp 43–50
45. Herzhaft B, Dalmazzone C (2001) Gas hydrate formation in drilling MUD characterized with DSC technique. In: SPE Annu. tech. conf. exhib. New Orleans, Louisiana, pp 575–584
46. Zhao X, Qiu Z, Zhou G, Huang W (2015) Synergism of thermodynamic hydrate inhibitors on the performance of poly (vinyl pyrrolidone) in deepwater drilling fluid. J Nat Gas Sci Eng 23:47–54
47. Gum X, Gum G, Gupta P, Nair VC, Sangwai JS (2019) Phase equilibrium of methane hydrate in aqueous solutions of polyacrylamide, xanthan gum, and guar gum. J Chem Eng Data 64:1650–1661

48. Wang S, Yuan C, Zhang C, Chen L, Liu J (2017) Rheological properties with tempera-
    ture response characteristics and a mechanism of solid-free polymer drilling fluid at low
    temperatures. Appl Sci 7:18
49. Wang R, Sun H, Shi X, Xu X, Zhang L, Zhang Z (2019) Fundamental investigation of the effects
    of modified starch, carboxymethylcellulose sodium, and xanthan gum on hydrate formation
    under different driving forces. Energies 12:2026
50. Koh C, Sloan D, Sum A (2010) Natural gas hydrates in flow assurance. Gulf Professional
    Publishing

Chapter 2
Fundamentals of Hydrates and Drilling Mud

This chapter is structured to provide basic principles and general processes on hydrates formation behavior and drilling mud properties. The kinetics and thermodynamic properties of hydrate formation are discussed to reflect the aspects of hydrate formation properties that cold be controlled in drilling mud. Also, the types of hydrate drilling mud systems are discussed to provide knowledge on their application for hydrate control during drilling.

2.1 Fundamentals of Hydrates

This section briefly discusses the fundamentals of gas hydrates needed for every drilling mud engineer. The definition of hydrate, its structure, and formation conditions are described to understand the effect of hydrate formation in mud as discussed in the preceding chapters.

2.1.1 Gas Hydrate and Its Formation Process

Gas hydrates are crystalline compounds that are formed when gas molecules are trapped in water molecules at low temperatures and high or adequate pressure conditions [1, 2]. When hydrates are formed, the water molecules do not have any chemical reactions with the gas, it is purely a physical reaction. With the gas and water molecules bonded by Van der Waals forces of attractions [3]. Hydrate formation is not limited to gases, some liquids such as tetrahydrofuran (THF) and cyclopentane forms hydrates under low-temperature conditions. When drilling oil and gas reservoirs or hydrate sediments, the common gases encountered are methane, ethane, propane, carbon dioxide, and other natural gas components [4]. The presence of these gases

B. Lal et al., *Hydrate Control in Drilling Mud*,
SpringerBriefs in Petroleum Geoscience & Engineering,
https://doi.org/10.1007/978-3-030-94130-7_2

shows that only structure I and structure II hydrates can be encountered while drilling hydrate-related wells. The types of gas hydrate structures are; structure I (sI), structure II (sII), and structure H (sH). The structure of gas hydrate is controlled by the guest molecule size and shape [5–7]. It also determines their thermodynamic or phase behavior conditions via controlling the guest or gas molecule cage occupancy. The phase behavior and cage occupancy are unit parameters that regulate the hydrate enthalpy of dissociation (H_d) [8]. The enthalpy of hydrate formation is very important to understand the effect of drilling mud on hydrate dissociation, this is because the dissociation behavior of hydrate in-situ rock could be affected the drilling mud thermal properties.

2.1.2 Hydrate Formation Process

The formation of hydrate is possible in any system consisting of low temperature, high pressure, gas, and water. During the drilling of oil and gas wells, hydrocarbons may flow into the wellbore via a kick [9]. The hydrocarbon released into the wellbore becomes guest molecules that possess a high hydrate formation risk in the system. On the other hand, when drilling hydrate sediments, the gas in the sediments as drilling rock cuttings are hydrate-bearing and must be transported out of the well safely to prevent hydrate reformation and instability. Thus, the presence of hydrocarbons and drilling mud in the wellbore possess a hydrate formation threat that requires the knowledge on gas hydrate formation process. Just like any crystallization process, gas hydrate formation involves phase changes initiated by a nucleation process, then followed by a further crystal growth process. The hydrate growth process is terminated by mass transfer or thermodynamic conditions limitations [8].

The formation of crystalline hydrates for the combination of water and gas molecules begins with a nucleation process [10]. This process causes the emergence of a new solid phase from the water and gas. This stage of hydrate formation is purely microscopic and involves the fusion or aggregation of very small particles to achieve a critical size. The time taken to achieve the critical particle size is known as the induction or nucleation time [11]. Knowing this time is very critical to prevent hydrate plugs while drilling. Practically the induction time denotes the time for the hydrate inhibitive drilling mud to fail in managing hydrate formation while drilling [12, 13]. Interestingly, the induction time is a stochastic phenomenon and could last for several periods of time, ranging from seconds to days. The duration of the induction depends on the properties and operation of the system. One most challenging task by researchers is the ability to accurately measure the hydrate nucleation time, owing to its stochastic nature. The most important parameters that control or govern the nucleation process is drilling force or subcooling, critical size, and energy barrier [14]. In the case of drilling, agitation or vibration from the drilling assembly can provide suitable or favorable interfacial gas–liquid contact that could enhance the nucleation process. Also, some drilling mud solid additives such as barites and bentonites coupled with the uncased rough rock surfaces can severe as hydrate nucleation sites to enhance hydrate nucleation or formation while drilling [15]. It must be stated that the absence of these factors would not prevent hydrate formation but might just delay or prolongs hydrate nucleation. Hydrate growth is next after hydrate nucleation. While hydrate nucleation is stochastic, its growth process is rapid and can be predicted. Factors such as pressure, vibrations, driving force, and system temperature control the hydrate growth process [16]. During the hydrate growth stage, the critical nuclei aggregate to form large crystals in the form of plugs. In this stage, the plug formation is stopped via mass transfer limitation.

2.1.3   Natural Occurring Hydrates Sediments

In the arctic and tropical regions, there is a vast region of permafrost and marine environment consisting of temperature conditions favorable for hydrate formation [17, 18]. These permafrost and marine environments are proven to contain huge amounts of gas hydrates, which is believed to be twice or more than the current fossil fuel in the world. About 10^{16} m^3 of methane gas is trapped as hydrates in these locations [19]. These huge amounts of clean natural gas could sever as future game-changing energy for the world if carefully exploited. Most of the hydrates discovered in nature are offshore although there are a few hydrates deposits found on land (permafrost) [20]. Geologically, gas hydrates continue to form in the earth, with more and more research for countries, researchers, and industries alike to extensively understand their drilling and production technologies. These studies are useful as the drilling of hydrate sediments poses a great risk which requires a deep understanding and breakthrough to successfully drilling those sediments.

Methane hydrates are located in the marine environment at depths ranging from 100 to 500 m, owing to the presence of suitable thermodynamic conditions. They are stable at such depths with some evidence of hydrate present at 2000 m [21]. These hydrates are known as high-temperature gas hydrate deposits, which are mostly encountered in deepsea drilling activities for conventional wells. On the other hand, hydrates are stable at 500–2000 m in the permafrost because of the lower freezing point existing in the arctic locations [21]. The temperature of the sediment and type of hydrate former (gas) presence defines the gas hydrate stability zone (GHSZ). Pressure plays a negligible role since the pressure is naturally defined by the sum of the hydrostatic and lithostatic pressure. In places such as the Gulf of Mexico, at a pressure equivalent of 2500 m, the base GHSZ will occur at about 294.15 K for pure methane hydrates, about 296.15 K for a typical mixture (methane 93%, ethane 4%, propane 1%, and other higher hydrocarbon) and 301.15 K for a possible mixture of methane 62%, ethane 9%, propane 23% and other hydrocarbons [22]. These differences in the temperature cause major shifts in the depths to the base of the GHSZ.

2.1.4 Drilling Hydrates Sediments

The formation of hydrates in the drilling mud and the dissociation of in-situ hydrates within drilled hydrate zones and sediment are the main issues that must be managed for safe operation and production in hydrocarbon development [23]. Unlike conventional drilling operations, the drilling of gas hydrate sediments is challenged with poor data and information. This poses a high drilling risk and needs much attention and caution in practical operation. The drilling of hydrate sediments presents challenges such as; plugging of borehole due to gas hydrate formation, blowout because of sudden gas hydrate dissociation, slope failure risk due to sudden gas hydrate dissociation, and wellbore stability problems and wellbore collapse risks due to the loose sediments after gas hydrate dissociation [15].

More details and practical hydrate sediment drilling operations and purposes are discussed in the last chapter of this book. However, it is believed that the choice of drilling mud could high prevent the above-stated challenges in hydrate sediment drilling activities [24]. Drilling mud highly affects the three main factors needed to achieve a successful drilling process (i.e. wellbore support, good wellbore cleaning, and cuttings transportation). Herein we focused on the experimental data and information on hydrate drilling mud needed to usefully drill hydrate sediments.

2.2 Fundamentals of Hydrates

Drilling muds are fluids generally used to bore holes (wells) to produce oil and gas (in the context of this study). Drilling muds are mainly used for wellbore control while drilling. Due to their fluid nature, density variations allow pressure counteractions while drilling at drilling depths. Drilling mud has different functions and performances depending on the type of formation being drilling. Thus, the drilling of hydrate sediments varies from conventional wells. The difference between hydrate sediment drilling muds and conventional oil well drilling muds is summarized in Table 2.1.

Table 2.1 Difference between hydrate sediment and conventional drilling mud

Parameters	Hydrate sediment drilling mud	Conventional well drilling mud
Temperature	Low-temperature mud	Moderate to high-temperature mud
Pressure	Relatively low-pressure withstanding muds (density related)	High pressure withstanding muds (density related)
Gas	Gas stability mud	No gas
Hydrates	Hydrate management and transportations	Absence of hydrate formation

2.2.1 Functions and Performance of Hydrate Drilling Mud

Drilling muds formulation consists of different additives mixed to provide specific or multiply functions while drilling. The most common and known types or categories of drilling mud additives are viscosifiers, weighting agents, fluid loss agents, pH/alkalinity control agents, hydrate inhibitors, shale inhibitors, corrosion inhibitors, etc. More details on drilling mud additives and their function can be found in Gray's Book [25]. The performance and functions of drilling mud are dependent on the intended type of rock to be drilled. Well properties such as pressure, temperature, cost, hydrate risk, and other complexities are the main factors considered when selecting and formulating drilling muds. The performance of drilling mud is controlled by the drilling assembly employed for drilling; this demands the altering of the mud properties to provide suitable hydraulics for the well that is being drilid.

The major functions of drilling muds are shown in Fig. 2.1. They are used to regulate pressure in the wellbore. Well pressures are controlled by drilling mud density variation. The mud pressures are estimated based on the fundamental knowledge of hydrostatic pressure. Aside from primary well control drilled cuttings transportation from the wellbore is another important function of drilling mud [26, 27]. The performance of drilling mud to transport cuttings from the well is controlled by the mud's density, viscoelastic properties, velocity, and other factors. These functions of the mud also ensure efficient well-cleaning for effective penetrations [27]. The drilling mud provides suitable properties that stabilizers and maintain the wellbore being drilling. In this regard, the mud fluid loss properties, filter cake quality, and pressure amongst others are designed and altered to maintain the wellbore. Hydrate

Fig. 2.1 Functions of drilling mud

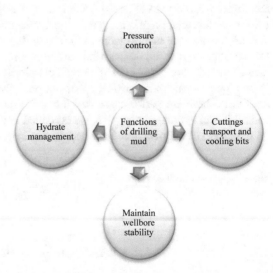

management is also am important function of drilling muds. They are used to prevent hydrate dissociation and plug formation in the wellbore and other frilling assembly facilities.

2.2.2 Types of Drilling Mud

The type of drilling mud used for any drilling operation is determined by the type of rock to be drilled or being drilled. However, other factors such as cost, environmental concerns, type, and quality of the based fluid to be used dictates the choice of mud being used. Also, the task-specific and performance of the muds could determine the type of mud to be used. There are main types of drilling mud on the classification of their base fluid composition. The three types of drilling mud are water-based drilling mud, oil-based drilling mud, and air-based drilling mud systems (see Fig. 2.2) [28]. In most drilling operations around the world, water-based mud systems are employed, preferably because of cost and environmental reasons. Oil-based muds are environmentally prohibitive and costly, though they have promising properties than water-based muds. The use of air-based muds is limited to drilling depleted zones and formations or reservoirs with very low pressured zones. Thus, air-based muds are rare in most drilling operations. The fundamental establishment of these muds is that water or brine solutions are the main phase in water-based muds, while in oil-based muds, the drilling additives are suspended in the oil continues phase.

Mostly, just having the drilling mud alone is not enough to sustain and provide the required properties needed to drilling wellbore. To enhance the mud properties to attain suitable functions, drilling additives such as fluid loss, viscosities, weighing and inhibitive additives are used. Details on the available type of drilling additives used in the formulation of drilling muds are tabulated in Figs. 2.3 and 2.4. In Figs. 2.3 and 2.4, the hydrate management fluid falls within the group of inhibitive fluids. In this section, we focused on the water and oil-based muds used for hydrate management. Thus, further types of drilling mud classification are considered with regards to hydrates management or drilling operations. Three types of hydrate muds are discussed, these are performance-based hydrate muds, water-based hydrate muds, and oil-based hydrate muds. Also, we classified and discussed hydrate muds based on the group of additives used for hydrate management in their base fluid system.

Fig. 2.2 Types of drilling mud

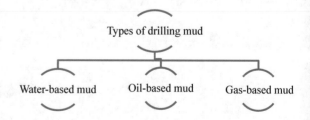

Types of drilling mud

Water-based mud   Oil-based mud   Gas-based mud

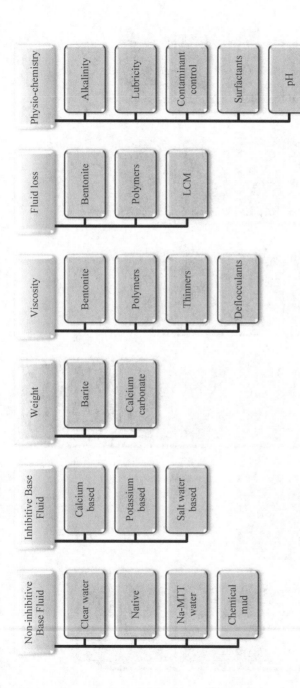

Fig. 2.3 Additives for water-based mud formulation[29]

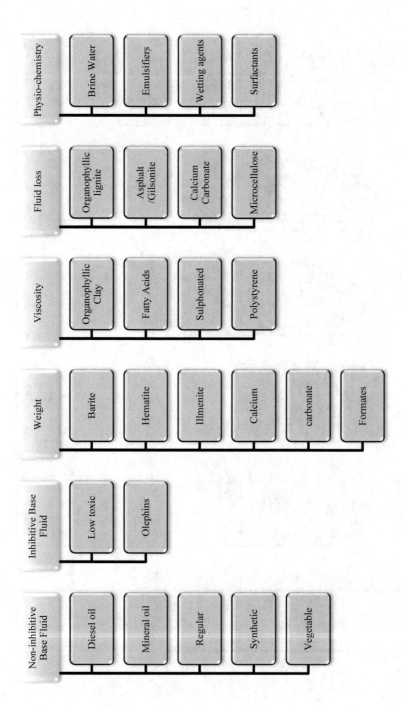

Fig. 2.4 Additives for oil-based mud formulation[29]

For water-based hydrate muds, we have the alcohol-based hydrate muds, polymer-based hydrate muds, fatty acids-based hydrate muds, polyether-based hydrate muds, nano-based hydrate muds, and ionic liquid-based hydrate muds. On the other hand, the oil-based hydrate muds are diesel-based muds and mineral oil-based muds.

2.2.3 Performance-Based Hydrate Muds

Performance-based hydrate muds are drilling mud that either affect the thermodynamic or kinetic behavior of hydrate formation and dissociation, as shown in Fig. 2.5. The kinetics could be during hydrate formation of dissociation. In hash deep offshore drilling operations TIMs and FKIMs are used to manage hydrate formation in the drilling mud. On the other hand, the drilling of hydrate sediments requires the use of FKIMs and DKIMs. The use of TIMs in drilling hydrate sediments may cause severe wellbore instability due to the dissociation of the in-situ hydrate by the mud filtrate. However, it must be stated that depending on the drilling mud additives used hydrate drilling mud could provide one, or two, and all three functions on hydrate formation.

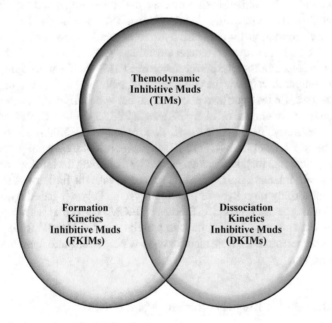

Fig. 2.5 Performance-based hydrate muds

2.2.3.1   Thermodynamic Inhibitive Muds (TIMs)

Thermodynamic inhibitive muds are drilling muds that affect the hydrate thermodynamic conditions (pressure and temperature). These muds generally shift the hydrate phase boundary conditions to lower temperatures and high pressures regions. Thus, reducing the hydrate risk zone. They are mostly used in the deepsea drilling operation to prevent hydrate plugs and formation in the chokelines and riser [30–32]. Hydrate studies involve thermodynamic studies that deal with gathering the incipient equilibrium hydrate formation data and developing predictive methods for the calculation of phase equilibria behavior. The incipient formation conditions refer to the situation in which an infinitesimal amount of the hydrate phase is present in equilibrium with fluid phases. Knowledge of the equilibrium hydrate-forming conditions is necessary for the rational and economic design of processes in the chemical, oil, gas, and hydrate sediment well design. Alcohol and salt-based muds are typical thermodynamic inhibitive muds.

2.2.3.2   Formation Kinetics Inhibitive Muds (FKIMs)

Formation Kinetics Inhibitive Muds are a class of drilling mud that delays the formation of kinetics hydrate in drilling mud systems. The main goals of FKIMs are to inhibit hydrate formation by delaying hydrate nucleation time or induction time, slow down the hydrate growth rate, and ensure less gas consumption into hydrates. Important factors that affect the performance of FKIMs are the subcooling temperature (ΔT = temperature below the equilibrium temperature) in the system, and pressure. Most often, FKIMs perform poorly at higher subcooling and high pressures. The subcooling temperature of many FKIMs is 282.15–283.15 K, indicating limitations in deep waters field applications. Two main hydrate inhibition mechanisms are suggested for KHIs [33, 34]. (1) FKIMs prevents hydrate nuclei particles from reaching a critical size for spontaneous growth by disturbing the local water structure via hydrophobic interactions. (2) FKIMs reduce or prevent further hydrate growth process by absorbing onto hydrate growing crystals. Polymer-based hydrate muds are a traditional type of FKIMs [35]. Studies on FKIMs performances are completely time-dependent and could last for days to months with the hydrate formation zone. This inhibition performance is needed for deep-water drilling and hydrate sediments drilling operation.

2.2.3.3   Dissociation Kinetics Inhibitive Muds (DKIMs)

Dissociation Kinetics Inhibitive Muds (DKIMs) are drilling muds systems that prevents or reduces the dissociation or melting of hydrate from hydrate sediments. This class of inhibitors do not affect the hydrate phase behavior like TIMs, however, their principle is opposite to FKIMs. The main focus of these muds is to inhibit hydrate dissociation by delaying the hydrate dissociation time, slowing down the hydrate dissociation rate, and reducing the amount of gas released from hydrates. Important factors that affect the performance of FKIMs are their disturbance on the hydrate phase boundary condition.

2.2.4 Additive-Based Hydrate Muds

Additive-based hydrate muds are the classification of hydrate drilling muds based on the type of additives used (see Fig. 2.6). The type of muds could exhibit one, two, or all the performance-based inhibition abilities. The types of additive-based hydrate muds discussed and tested in literature are present herein. Additive-based muds are either water-based muds systems or oil-based muds systems.

2.2.4.1 Water-Based Hydrate Muds

Water-based hydrate muds are the most widely used. They comprise a mixture of water and reactive solids, inert solids, functional chemicals, and sometimes non-aqueous liquids as used in the industry. However, the presence of hydrate managing additives makes them unique for hydrate-related drilling operations. Figure 2.6 presents the available water-based hydrate muds systems in literature.

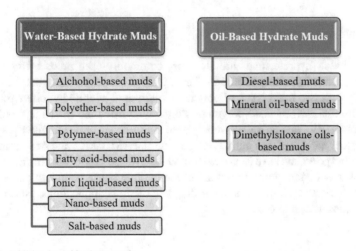

Fig. 2.6 Additive-based hydrate muds

Alcohol-Based Hydrate Muds

Alcohol-based hydrate muds are water-based muds formulated with alcohol-based hydrate inhibitors. These alcohols are generally thermodynamic gas hydrate inhibitors, consisting of MeOH, and EtOH. Glycols also fall in the same category, common additives such as ethylene glycol, diethylene glycols, TEG, and glycerol [15, 32]. These inhibitors mainly act by shifting the hydrate phase behavior condition to a less hydrate-prone zone. However, these inhibitors are challenged by environmental concerns. However, they're mostly used deep-offshore operations. Their thermodynamic inhibition effect discourages their use for drilling hydrate sediments. This is because they could potentially release a huge amount of gas into the wellbore.

Polyether-Based Hydrate Muds

The use of polyether hydrate additives in drilling is reported to provide good hydrate management potentials. Polyether such as polyglycol (PEG), ethyl butyrate, n-propyl propionate, n-butyl butyrate, and n-amyl butyrate have been used to formulated polyether-based hydrate muds. The use of mixed di-propylene glycol methyl ether base and xanthan biopolymer (43%) known as XANVIS-L is a hybrid ether-polymer-based hydrate mud for hydrate management. The class of inhibitors could either function as FKIM or TIMs.

Polymer-Based Hydrate Muds

Polymer-based hydrate muds are also very commonly used muds owing to their hydrate kinetics inhibition potentials. These classes of hydrate muds are formulated basically with two main types of polymers: traditional hydrate inhibitive polymers and traditional drilling mud polymers. Polyvinylpyrrolidone (PVP) and polyvinyl caprolactam (PVCap) are the commonly used traditional hydrate inhibitive polymers. On the other hand, xanthan gum, PAC, guar gum, PAM, CMC, and other traditional drilling mud polymers have been used to study as hydrate suppressors in drilling mud. These classes of inhibitors do not affect the hydrate phase boundary curve. Thus, as useful for both hydrate sediment drilling and offshore applications, especially for short periods of operations.

Fatty Acid-Based Hydrate Muds

Fatty acids are generally compounds consisting of carboxylic acid with a long aliphatic chain, which is either saturated or unsaturated. Their potentials as kinetics hydrate inhibitors have encouraged their use in drilling mud for hydrate management. Hydrate drilling muds formulated from lecithin, ethyl butyrate, n-propyl propionate, n-butyl butyrate, and n-amyl butyrate are the main fatty acid-based muds in literature.

Ionic Liquid-Based Hydrate Muds

Ionic liquids are organic salts that have been introduced as novel gas hydrate inhibitors [36–39]. They are known to have good thermal stability, strong electrostatic forces of attraction, relatively green, and relatively cheap, and soluble in water [40]. Recent studies have shown that ionic liquids can maintain the shear-thinning properties of water-base muds. Mud formulations using ionic liquids are known as ionic liquid-based muds [28, 41]. This class of muds is believed to provide a dual or triple hydrate inhibition mechanism. They could also be used in conventional drilling owing to the thermal and rheological properties enhancement potentials. Since they are salts, there is a huge ionic liquid database that could be tested and used. They are designer compounds because they allow easy cation and anion pairing with the inclusion of functional groups for a specific task. However, the use of ionic liquids for hydrate management in drilling mud is still at its initial stage with much ongoing research. Generally, imidazolium and ammonium-based ionic liquids muds have been studied in literature with Cl, Br, BF_4 halides [36, 38, 42, 43].

Nano-based Hydrate Muds

Nanoparticles or ultrafine particles are usually defined as a particle of matter that is between 1 and 100 nm (nm) in diameter. Nana-based drilling fluid has been tested for conventional drilling mud systems. However, in hydrate drilling carbon dot-based nana fluids have been studied by far.

Salt-Based Muds

These classes of drilling muds are mainly formulated with salts and exhibit a thermo-dynamic hydrate inhibition effect. Salts-based muds are believed to be the most used muds system in preventing hydrate during offshore drilling of 2286 m. Salts hardly have kinetic hydrate inhibition, their thermodynamic inhibition nature prohibits their excessive use in drilling hydrate sediments. The most used salt-based muds are NaCl/polymer-based muds.

2.2.4.2 Oil-Based Hydrate Muds

The oil-based hydrate muds are mainly formulated using diesel. Mineral oil and Dimethyl siloxane oils. These oils each have a varying effect on hydrate formation. The use of diesel oil is common, which the use of mineral and Dimethyl siloxane oils is more related to academic literature.

2.2.5   Additives Used for Hydrate Management

Herein we present details of hydrate additives in drilling mud formulation and their effect on hydrate inhibition. The information in Table 2.2 provided basic information and knowledge to select additives for hydrate mud formulations.

2.2.6   Rheology of Drilling Mud

The rheology of drilling mud deals with the study and evaluation of the internal structural changes of the mud when it undergoes shearing through the application of force on it. In other words, the study of the deformation and flow behavior of drilling mud characterizes their rheological properties. In drilling mud, rheological studies various equipment are used to evaluate the mud deformation and flow behavior in a manner to mimic the muds flow behavior while drilling. Mostly, pressure, temperature, and shearing variations are used to test the mud rheological properties and hydraulics. This section briefly discusses the important drilling mud rheological properties mostly studied on hydrate management drilling mud systems. Rheological properties such as viscosity, plastic viscosity, apparent viscosity, shear rate, shear stress, yield stress, etc. are discussed.

2.2.6.1   Shear Stress, Shear Rate, and Viscosity

Shear rate and stress are the basic conditions to initial flow in any system. The shear stress is representing the amount of force applied to a fluid to cause the fluid to shear at a respective rate. In drilling the speed or velocity of pumping the mud and the amount of pressure need to cause the fluid to flow through to the surface defines the shear rate and shear stress behavior [25].

In practice, viscosity is the most used rheological property in every field. It simply defines the ability of a material to oppose flow, which is commonly referred to as the ticking or thinning nature of the fluid. In scientific writing, η is used to denote viscosity, which is mathematically represented as the ratio of shear stress and shear rate, as shown in Eq. 2.1.

$$\eta = \frac{\tau}{\gamma} \tag{2.1}$$

where τ is the shear stress, γ is the shear rate.

Aside, temperature, pressure also affects viscosity to some extent. However, the effect of pressure on viscosity is negligible. The effect of temperature is very significant on viscosity and thus needs to be well understood. While drilling oil and gas wells, temperature variations are very important to monitor and evaluate their effect

Table 2.2 Summary of hydrate management drilling additives and their hydrate inhibition properties

Additive	Formula	MW (gmol^{-1})	Density (gcm^{-3})	Classification	Hydrate performance	Remarks
Methanol	CH_3OH	32.04	0.792	Alcohol	THI	Preferred for offshore drilling with deep mud lines
Ethanol	C_2H_6O	46.07	0.789	Alcohol	THI	Preferred for offshore drilling with deep mud lines
Ethylene glycol	$C_2H_6O_2$	62.068	1.113	Glycol	THI	Preferred for offshore drilling with deep mud lines. Has been reported as a potential hydrate dissociation inhibitor
Triethylene glycol	$C_6H_{14}O_4$	150.174	1.126	Glycol	THI	Preferred for offshore drilling with deep mud lines
Diethylene glycol	$C_4H_{10}O_3$	106.12	1.118	Glycol	THI	Preferred for offshore drilling with deep mud lines
Polyethylene glycol	$C_{2n}H_{4n+2}O_{n+1}$	44.05n + 18.02	1.125	Glycol	THI	Preferred for offshore drilling with deep mud lines
Propylene glycol	$C_3H_8O_2$	76.095	1.036	Glycol	THI	Preferred for offshore drilling with deep mud lines
glycerol	$C_3H_8O_3$	92.094	1.261	Polyol	THI	Preferred for offshore drilling with deep mud lines
Hexamethyl Disiloxane	$(CH_3)_6Si_2O$	162.37752	0.75–0.77	DO	NA	Has suitable electrical and thermal conductivity for drilling hydrate sediments

(continued)

Table 2.2 (continued)

Additive	Formula	MW (gmol^{-1})	Density (gcm^{-3})	Classification	Hydrate performance	Remarks
Silicon fluid	$(CH_3)_8Si_3O$	Wide range	0.76- 0.96	DO	NA	Has suitable electrical and thermal conductivity for drilling hydrate sediments
n-Propyl propionate	$C_6H_{12}O_2$	116.160	0.833	Esters	NA	Has suitable electrical and thermal conductivity for drilling hydrate sediments
Ethyl butyrate	$C_6H_{12}O_2$	116.160	0.879	Esters	NA	Has suitable electrical and thermal conductivity for drilling hydrate sediments
n-butyl butyrate	$C_8H_{16}O_2$	144.214	0.869	Esters	NA	Has suitable electrical and thermal conductivity for drilling hydrate sediments
n-amyl butyrate	$C_9H_{18}O_2$	158.24	0.860	Esters	NA	Has suitable electrical and thermal conductivity for drilling hydrate sediments
Sulfomethylated Phenolic Resin	NA	NA	NA	Polymer	KHI	Could be used for any hydrate related drilling operation
Sodium metasilicate	Na_2SiO_3	122.06	2.4	Electrolyte	THI	Preferred for offshore drilling with deep mud lines
Potassium silicate	$K_2Si_2O_5$	154.279	1.39	Electrolyte	THI	Preferred for offshore drilling with deep mud lines

Additive	Formula	MW (gmol^{-1})	Density (gcm^{-3})	Classification	Hydrate performance	Remarks
Potassium bromide	KBr	119.002	2.74	Electrolyte	THI	Preferred for offshore drilling with deep mud lines

(continued)

Table 2.2 (continued)

Additive	Formula	MW (gmol^{-1})	Density (gcm^{-3})	Classification	Hydrate performance	Remarks
Barium chloride	BaCl$_2$	208.23	3.856	Electrolyte	THI	Preferred for offshore drilling with deep mud lines
Potassium chloride	KCl	74.551	1.98	Electrolyte	THI	Preferred for offshore drilling with deep mud lines
Sodium chloride	NaCl	58.443	2.17	Electrolyte	THI	Preferred for offshore drilling with deep mud lines
Sodium bromide	NaBr	102.894	3.21	Electrolyte	THI	Preferred for offshore drilling with deep mud lines
Strontium chloride	SrCl$_2$	158.53	3.052	Electrolyte	THI	Preferred for offshore drilling with deep mud lines
Potassium formate	CHKO$_2$	84.115	1.908	Electrolyte	THI	Preferred for offshore drilling with deep mud lines
Cesium formate	CHCsO$_2$	177.92	2.4	Electrolyte	THI	Preferred for offshore drilling with deep mud lines
Sodium carbonate	Na$_2$CO$_3$	105.989	2.54	Electrolyte	THI	Preferred for offshore drilling with deep mud lines
Trisodium phosphate	Na$_3$PO$_4$	163.94	2.536	Electrolyte	THI	Preferred for offshore drilling with deep mud lines

(continued)

Table 2.2 (continued)

Additive	Formula	MW (gmol^{-1})	Density (gcm^{-3})	Classification	Hydrate performance	Remarks
Magnesium chloride	$MgCl_2$	95.211	2.32	Electrolyte	THI	Preferred for offshore drilling with deep mud lines
Sodium sulfate	Na_2SO_4	142.04	2.664	Electrolyte	THI	Preferred for offshore drilling with deep mud lines
Calcium chloride	$CaCl_2$	110.98	2.15	Electrolyte	THI	Preferred for offshore drilling with deep mud lines
Sodium hydroxide	$NaOH$	39,997	2.13	Electrolyte	THP	Might minimize hydrate dissociation
Ammonium calcium nitrate	H_4CaN_4O	244.13	1.0–1.1	Electrolyte	THI	Preferred for offshore drilling with deep mud lines
Lignosulfonate	$C_{20}H_{24}Na_2O_{10}S_2$	534.5	0.5	Polymers	THP/KHP	Might minimize hydrate dissociation
Bentonite	$Al_2H_2Na_2O_{13}Si_4$	422.29	2.2–2.8	Clay	THP/KHP	Might minimize hydrate dissociation
Polyvinylpyrrolidone	$(C_6H_9NO)_n$	2,500 – 2,500,000	1.2	Polymer	KHI	Could be used for any hydrate-related drilling
Carboxymethyl cellulose	$C_8H_{15}NaO_8$	variable	Variable	Polymer	KHI	Could be used for any hydrate-related drilling
Xanthan gum	$(C_{36}H_{58}O_{29}P_2)_n$	variable	Variable	Polymer	KHI	Could be used for any hydrate-related drilling
Polyacrylamide	$(C_3H_5NO)_n$	Variable	Variable	Polymer	THP/KHP	Might minimize hydrate dissociation

(continued)

Table 2.2 (continued)

Additive	Formula	MW (gmol^{-1})	Density (gcm^{-3})	Classification	Hydrate performance	Remarks
Barite	$BaSO_4$	233.38	4.48	Dolomite	Unknown	Not applicable
Lipase	$C_{45}H_{69}NO_8$	752	1.2–1.35	Lipids	KHI	Could be used for any hydrate related drilling operation
Lecithin	$C_{35}H_{66}NO_7P$	643.9	1.3 ± 0.1	Lipids	KHI	Could be used for any hydrate related drilling operation

Additive	Formula	MW (gmol^{-1})	Density (gcm^{-3})	Classification	Hydrate performance	Remarks
Polyanionic cellulose	$[C_6H_7O_2(OH)_2CH_2COONa]_n$	Variable	1.43–5.71	Polymer	KHI	Could be used for any hydrate related drilling operation
Guar gum	$C_{10}H_{14}N_5Na_2O_{12}P_3$	50,000–8,000,000	0.8–1	Polymer	KHI	Could be used for any hydrate related drilling operation
Starch	$C_{12}H_{25}NO_{11}$	359.33	1.5	Polymer	KHI	Could be used for any hydrate related drilling operation
2-(Dimethylamino) ethyl methacrylate	$C_8H_{15}NO_2$	157.21	0.933	Polymer	KHI	Could be used for any hydrate related drilling operation
1-decyl-3-methylimidazolium tetrafluoroborate	$C_{14}H_{27}BF_4N_2$	310.18	1.444	Ionic liquid	THI/KHI	Could be used for any hydrate related drilling operation

(continued)

Table 2.2 (continued)

Additive	Formula	MW (gmol^{-1})	Density (gcm^{-3})	Classification	Hydrate performance	Remarks
1-Ethyl-3-methylimidazolium chloride	$C_6H_{11}ClN_2$	146.62	1.186	Ionic liquid	THI/KHI	Could be used for any hydrate-related drilling operation
Tetramethylammonium chloride	$C_4H_{12}ClN$ or $(CH_3)_4NCl$	109.6	1.17	QAS	THI/KHI	Could be used for any hydrate-related drilling operation
Tetramethylammonium bromide	$(CH_3)_4 N(Br)$	154.05	1.56	QAS	THI/KHI	Could be used for any hydrate-related drilling operation
Tetrapropylammonium bromide	$(CH_3CH_2CH_2)_4 N(Br)$	266.26	1.15	QAS	THI/KHI	Could be used for any hydrate-related drilling operation
1-Butyl-3-methylimidazolium chloride	$C_8H_{15}ClN_2$	174.67	1.086	Ionic liquid	THI/KHI	Could be used for any hydrate-related drilling operation
1-Butyl-3-methylimidazolium bromide	$C_8H_{15}BrN_2$	219.12	1.300	Ionic liquid	THI/KHI	Could be used for any hydrate-related drilling operation
1-Hexyl-3-methylimidazolium chloride	$C_{10}H_{19}ClN_2$	202.72	1.0337	Ionic liquid	THI/KHI	Could be used for any hydrate-related drilling operation
1-Methyl-3-octylimidazolium chloride	$C_{12}H_{23}ClN_2$	230.78	1.01	Ionic liquid	THI/KHI	Could be used for any hydrate-related drilling operation

Dimethylsiloxane oil (DO)

on the rheology of the mud. In hydrate wells, the effect of low temperature on the rheology of the mud is most needed to comprehend. The viscosity of drilling mud is usually not constant when drilling a well, hence, the mud viscosity is constantly monitored while drilling. Plastic viscosity (*PV*) is another type of viscosity estimated in the Bingham plastic model. It is used to describe the mechanical friction viscosity of drilling as a function of the concentration of solids in the mud [44].

2.2.6.2   Yield Stress

The study of drilling mud yield stress is important to understand the amount of force needed to initial flow while drilling. The yield stress also has the ability to indirectly describe to gelling nature of the field. The yield stress in drilling mud systems is the evaluation of the electrochemical and/or attractive forces in the mud. The mud's electrochemical properties are due to the solid's concentration in the mud, which can be affected by drilling properties and chemical additives. The elastic or gelling properties are defined as the mud's behavior when any stress below its yield stress is applied. In this regard, it is considered that the mud of fluid has not undergone deformations. This property of the mud is very useful to understand the cutting suspension and carrying capacity. Another name for yield strass is yield point, mostly used in Bingham Plastic model estimation. The most accurate method to estimate yield stress is to use a rheometer, instead of the conventional viscometer.

2.2.6.3   Thixotropy and Shear Thinning

Thixotropy is the decrease in fluid viscosity with time, while the decrease in viscosity with increasing shear rate in shear thinning. The time-dependent nature (thixotropic) and shear thinning properties of drilling muds are due to the presence of polymers and clay viscosifiers used for their formulations. The shear-thinning nature of the drilling mud causes a low viscosity to pump the drilling mud through the drilling bits, while provided a relaxed viscosity to carry cuttings out of the wellbore. On the other hand, the thixotropy of drilling mud is reversible and can re-settle the mud microstructural properties at rest. This behavior describes the gelling ability of the mud under shut-in conditions. The gelling properties cause the cuttings to suspend in the muds. In simple viscometer measurement, the carrying capacity of the drilling mud is defined by the ratio of their YP/PV. The use of rheometric evaluations to test the gelling properties of the mud is more accurate. However, high gel or YP/PV could lead to high-pressure requirements, which can lead to poor wellbore cleaning efficiency. Another conventional technique to measure gel strength is the 10 s and 10 min shearing on a viscometer [45].

2.2.6.4 Newtonian Fluids and Non-Newtonian Fluids

Generally, fluids can be categorized as Newtonian or non-Newtonian based on their shear rate versus shear stress relationship (flow curve). Fluids exhibit a constant slope (viscosity) for a shear rate and shear stress is known as a Newtonian fluid. Newtonian fluids are usually water, brine solution, and oil. However, at a high shear rate, Newtonian fluids deviate to behavior like non-Newtonian.

For a consistency plot for a fluid, the absence of constant viscosity denotes the presence of non-Newtonian behavior. Implying that in non-Newtonian fluids the shear stress is not directly proportional to the shear rate. Non-Newtonian fluids are complex in nature and could exhibit different behavior with shear. In a non-Newtonian mode, the properties of the fluid could be shear dependant or independent. Also, the behavior of the fluid might be described by various properties on the solids in the fluids, this is known as a viscoelastic behavior. Generally, drilling muds are non-Newtonian fluids and mostly exhibit viscoelastic behavior [25, 44].

2.2.6.5 Viscoelasticity

The elastic and viscous behavior of fluids under deformation is known as their viscoelasticity. Unlike viscosity, which deals with the internal structure of the fluids, fluid elasticity deals with the mechanics of the solids in the muds by exhibiting the capacity of the mud to assumes its original state when strained. As mentioned earlier, drilling mud exhibits viscoelastic behavior, thus, their viscoelastic behavior can be evaluated using and viscoelastic equipment. From the fundamentals of physics, elastic bodies have the ability to store energy, and thus, are capable of withstanding strain. On the other hand, viscous fluids release energy when strain and do not restore to their initial properties. Therefore, in a viscoelastic fluid, like drilling mud, the fluid has the ability to store some amount of energy for its restorations and well loss the remaining energy for initial flow [46].

The viscoelastic behavior of fluids is evaluated by measuring their storage modulus (G'), loss modulus (G''), and complex viscosity (η^*). The storage modulus denotes the elastic property of the fluid, which is related to the capacity of the fluid to store energy when shear is applied. The loss modulus denotes the viscous property of the fluid to lose energy when sheared. The ratio of the loss modulus to the storage modulus is known as the damping factor (see Eq. 2.2). The liquid or solid behavior of the fluid can be determined from the damping factor [47].

$$\tan(\delta) = \frac{G''}{G'} 273.15\,K \leq \delta \leq 363.15K \tag{2.2}$$

From Eq. 2.2, a $\delta = 273.15$ K indicates an ideal elastic solid behavior, while $\delta = 363.15$ K denotes a viscous liquid. Therefore, a $\delta = 318.15$ K is the transition of the fluid from elastic to viscous, which reflects the yield stress of the fluid. During viscoelastic measurement, the complex viscosity ($\eta*$) is the ability of the fluid to oppose flow under the oscillatory mode, instead of shearing is the normal viscosity measurement [48].

References

1. Bavoh CB, Nashed O, Khan MS, Partoon B, Lal B, Sharif AM (2018) The impact of amino acids on methane hydrate phase boundary and formation kinetics. J Chem Thermodyn 117:48–53
2. Bavoh CB, Lal B, Keong LK, Jasamai MB, Idress MB (2016) Synergic kinetic inhibition effect of EMIM-Cl + PVP on CO_2 hydrate formation. In: Procedia Eng., pp 1232–1238
3. Bavoh CB, Lal B, Khan MS, Osei H, Ayuob M (2018) Combined inhibition effect of 1-ethyl-3-methy-limidazolium Chloride + Glycine on methane hydrate. J Phys Conf Ser 1123:012060
4. Bavoh CB, Lal B, Osei H, Sabil KM, Mukhtar H (2019) A review on the role of amino acids in gas hydrate inhibition, CO_2 capture and sequestration, and natural gas storage. J Nat Gas Sci Eng 64:52–71
5. Jeffery GA (1972) Pentagonal dedecahedral water structure in crystalline hydrates. Mat Res Bull 7:1259–1270
6. Jeffrey GA (1984) Hydrate inclusion compounds. J Incl Phenom 1:211–222
7. Ripmeester JA, Tse JS, Ratcliffe CI, Powell BM (1987) A new clathrate hydrate structure. Nature 325:135–136
8. Bavoh C.B., Lal B. KLK (2020) Introduction to gas hydrates. In: Chem. addit. gas hydrates. Green Energy and Technology. Springer, Cham, pp 1–20
9. Østergaard KK, Tohidi B, Danesh A, Todd AC (2000) Gas hydrates and offshore drilling. Predicting the hydrate free zone. Ann N Y Acad Sci 912:411–419
10. Kashchiev D, Firoozabadi A (2002) Nucleation of gas hydrates. J Cryst Growth 243:476–489
11. Sloan ED, Koh CA (2008) Clathrate hydrates of natural gases, 3rd edn. CRC Press Taylor & Francis, Boca Raton; London; New York
12. Kashchiev D, Firoozabadi A (2003) Induction time in crystallization of gas hydrates. J Cryst Growth 250:499–515
13. Nashed O, Sabil KM, Ismail L, Japper-Jaafar A, Lal B (2018) Mean induction time and isothermal kinetic analysis of methane hydrate formation in water and imidazolium based ionic liquid solutions. J Chem Thermodyn 117:147–154
14. Khurana M, Yin Z, Linga P (2017) A review of clathrate hydrate nucleation. ACS Sustain Chem Eng 5:11176–11203
15. Hale H, Dewan AKR (1990) Inhibition of gas hydrates in deepwater drilling. SPE Drill Eng 7:109–115
16. Dendy Sloan E, Koh CA (2008) Gas hydrates of natural gases, 3rd edn. CRC Press LLC, 2000 Corporate Blvd., N.W., Boca Raton, FL 33431, USA Orders from the USA and Canada (only) to CRC Press LLC, London; New York
17. Ruppel CD, Kessler JD (2017) The interaction of climate change and methane hydrates. Rev Geophys 55:126–168
18. Koh CA (2002) Towards a fundamental understanding of natural gas hydrates. Chem Soc Rev 31:157–167
19. Englezos P (1993) Clathrates hydrates. Ind Eng Chem Res 32:1251–1274
20. Kvenvolden KA (1988) Methane hydrate—a major reservoir of carbon in the shallow geosphere? Chem Geol 71:41–51
21. Merey Ş (2016) Drilling of gas hydrate reservoirs. J Nat Gas Sci Eng 35:1167–1179

22. Darley HCH, Gray GR (1983) Composition and properties of drilling and completion fluids, 5th edn. Gulf Professional Publishing, Houston, TX
23. Khabibullin T, Falcone G, Teodoriu C (2011) Drilling through gas-hydrate sediments: managing wellbore-stability risks. SPE Drill Complet 26:287–294
24. Jiang G, Ning F, Zhang L, Tu Y (2011) Effect of agents on hydrate formation and low-temperature rheology of polyalcohol drilling fluid. J Earth Sci 22:652–657
25. Caenn R, Darley HCH, Gray GR (2017) Drilling fluid components. In: Compos. prop. drill. complet. fluids. Elsevier Inc, pp 537–595
26. Kim NR, Ribeiro PR, Bonet EJ (2007) Study of hydrates in drilling operations: a review. Braz J Pet Gas 1:116–122
27. Paul Scott, Broussard P, Bland R, Growcock F, Freeman M (2015) Drilling Fluids. In: IADC Drill. Man., 12th Ed. p 66
28. Ofei TN, Bavoh CB, Rashidi AB (2017) Insight into ionic liquid as potential drilling mud additive for high temperature wells. J Mol Liq 242:931–939
29. Caenn R, Darley HCH, Gray GR (2017) The rheology of drilling fluids. In: Compos. prop. drill. complet. fluids. Elsevier Inc, pp 151–244
30. Hale AH, Dewan AKR (1990) Inhibition of gas hydrates in deepwater drilling. SPE Drill Eng 5(109–115):18638
31. Lai DT, Dzialowski AK (1989) Investigation of natural gas hydrates in various drilling fluids. In: SPE/IADC drill conf New Orleans, Louisiana 18637, pp 181–194
32. Yousif MH, Young DB (1993) Simple correlation to predict the hydrate point suppression in drilling fluids. In: Drill conf—proc, pp 287–294
33. Koh CA, Sloan ED, Sum AK, Wu DT (2011) Fundamentals and applications of gas hydrates. Annu Rev Chem Biomol Eng 2:237–257
34. Bavoh CB, Lal B, Osei H, Sabil KM, Mukhtar H (2019) A review on the role of amino acids in gas hydrate inhibition, CO_2 capture and sequestration, and natural gas storage. J Nat Gas Sci Eng 64:52–71
35. Lal B, Nashed O (2019) Chemical additives for gas hydrates
36. Bavoh CB, Ofei TN, Lal B, Sharif AM, Shahpin MHBA, Sundramoorthy JD (2019) Assessing the impact of an ionic liquid on NaCl/KCl/polymer water-based mud (WBM) for drilling gas hydrate-bearing sediments. J Mol Liq 294:111643
37. Bavoh CB, Lal B, Nashed O, Khan MS, Keong LK, Bustam MA (2016) COSMO-RS: an ionic liquid prescreening tool for gas hydrate mitigation. Chin J Chem Eng 11:1619–1624
38. Bavoh CB, Ofei TN, Lal B (2020) Investigating the potential cuttings transport behavior of ionic liquids in drilling mud in the presence of sII hydrates. Energy Fuels 34:2903–2915
39. Khan MS, Liew CS, Kurnia KA, Cornelius B, Lal B (2016) Application of COSMO-RS in investigating ionic liquid as thermodynamic hydrate inhibitor for methane hydrate. Procedia Eng 148:862–869
40. Dharaskar SA, Varma MN, Shende DZ, Yoo CK, Wasewar KL (2013) Synthesis, characterization and application of 1-butyl-3 methylimidazolium chloride as green material for extractive desulfurization of liquid fuel. Sci World J 2013:1–9
41. Bavoh CB, Ntow T, Lal B, Sharif AM, Shahpin MHBA, Sundramoorthy JD (2019) Assessing the impact of an ionic liquid on NaCl/KCl/polymer water-based mud (WBM) for drilling gas hydrate-bearing sediments. J Mol Liq 294:111643
42. Yuha YBM, Bavoh CB, Lal B, Broni-Bediako E (2020) Methane hydrate phase behaviour in EMIM-Cl water based mud (WBM): an experimental and modelling study. S Afr J Chem Eng 34:47–56
43. Bavoh CB, Yuha YB, Tay WH, Ofei TN, Lal B, Mukhtar H (2019) Experimental and modelling of the impact of quaternary ammonium salts/Ionic Liquid on the rheological and hydrate inhibition properties of Xanthan gum water-based muds for drilling gas hydrate-bearing rocks. J Pet Sci Eng 183:106468
44. API (2010) 13D—recommended practice on the rheology and hydraulics of oil-well drilling fluids
45. Barnes AH (1997) Thixotropy-a review. J Non-Newtonian Fluid Mech 70:1–33

46. Bui B (2012) Viscoelastic properties of drilling fluids. Annu Trans Nord Rheol Soc 20:33–47
47. Mezger T (2006) The rheology handbook—for users of rotational and oscillatory rheometers. Vincentz Verlag, Hannover-Germany
48. Gutierrez-Lemini D (2014) Engineering viscoelasticity. Springer, New York

Chapter 3
Testing Hydrate Drilling Fluid Properties

Drilling fluid properties are mostly measured and tuned to suit a specific function needed for drilling a particular drilling section and location [1]. Therefore, routine measurements are conducted to ensure that the drilling mud properties are suitable at every drilling section. In this chapter, the methods for testing drilling fluid properties described by API [2] are summarized here with further descriptions of the hydrate management properties testing in drilling muds.

3.1 Drilling Fluid Rheological Properties Testing Methods

3.1.1 Mud Density Measurement

In the field, a mud balance is used to measure the density of drilling mud systems. This property is a measure of the mass per unit volume of the mud. The density of the mud is also presented in specific gravity or SG (g/ml). The significance of measuring the density is to estimate the required hydrostatic pressure needed at any drilling depth for pressure control [1].

3.1.2 Mud Viscosity Measurement

As discussed in Chap. 2, viscosity is the common rheological property of drilling mud often measured. Common methods for measuring the viscosity of drilling mud on the field are the Marsh funnel and the use of a rotational viscometer. The gel strength of the mud is also determined using rotational viscometers. The use of viscometers to measure the viscosity of drilling muds is governed by estimating the

B. Lal et al., *Hydrate Control in Drilling Mud*,
SpringerBriefs in Petroleum Geoscience & Engineering,
https://doi.org/10.1007/978-3-030-94130-7_3

viscosity at varying shear rates [1]. The dial reading on viscometers are converted to shear rates with their respective shear stress values estimated from the viscometer constant estimation chart. The measured shear stress for the shear rate for 600 and 300 rpm is mostly used for estimating the parameters (plastic viscosity (PV) and yield point (YP)) of the Bingham plastic model. The plastic viscosity is estimated experimentally by subtracting the dial reading at 300-rpm dial reading from 600-rpm dial reading, as shown in Eq. 3.1. The yield point estimation is given in Eq. 3.2 as the subtracting of the PV from the dial reading at 300-rpm. It is of best practice to measure the viscosity of the mud at a suitable temperature that mimics the drilling operation. For hydrate sediments, the standard temperature for testing drilling ranges from -2 – 4 °C. while for conventional wells, the standard temperature ranges from 120 to 150 °C [1].

$$PV\,(mPa.s) = \theta_{600} - \theta_{300} \qquad\qquad (3.1)$$

$$YP\,(47.88Pa) = \theta_{300} - PV \qquad\qquad (3.2)$$

3.1.3 Gel Strengths and Filtration Test

The gel strength of the drilling mud systems is the minimum amount of stress needed to overcome the static period of the mud. The gel strength defines the gelling behavior of the mud under static conditions. The gel common method to measure the gel strength of drilling mud is to read the shear stress point on a viscometer at very low revolutions per minute (3 rpm) [1]. Prior to the testing, the mud must be put to rest. Mostly, the gel strength values are read for 10 s, 10 min, and 30 min. The measurement of the filtrate and mud cake thickness is very important for know how much the mud would lose liquids to the drilling formation. Filtration or fluid loss property of the mud defines the amount of fluid in the mud that would possibly escape through the deposited solids (mud cakes) of the drilling mud on the rock surface. The fluid loss property of drilling mud is mostly determined by collecting the filter volume of the mud through a standard filter paper over a period of 30-min in a filter press at ambient or high-pressure conditions. This test is very good in hydrate drilling mud because, if the filtrate has the ability to chemically inhibit hydrates, they may dissociate the hydrate in the rock. Which could lead to wellbore collapse and hydrate plug challenges. Similarly, the temperature of the filtrate could also cause hydrate dissociation of reformation in the wellbore. Thus, its suitable to use drilling mud systems that have minimal filter volume when drilling hydrate sediments or zones. In most cases, reducing the solid content of the mud, coupled with the use of effective filtration additives could provide quality mud cakes and filtration behavior [1].

3.1.4 Rheometric Testing

As mentioned earlier, the viscoelastic properties of drilling muds cannot be measured using the conventional viscometer [3, 4]. Thus, the use of rheometers provides suitable opportunities to evaluate the viscoelastic behavior of drilling muds. For hydrate testing purposes, the ability to test the viscoelastic properties of the mud in the presence of hydrate can provide significant knowledge on how to transport hydrates or gas in the drilling mud to the surface [5]. However, such studies are very limited in literature and need to be addressed. For accurate measurement, rheometric analysis for drilling should move from the viscometer to the use of rheometers. In the testing technique, the linear viscoelastic range (LVER) of the drilling mud is evaluated by applying small deformation to the mud sample [3]. Generally, viscoelastic tests are known as dynamic tests. Under the dynamic test conditions, a transient and oscillatory testing mode can be conducted. In the oscillatory test mode, frequency sweep, amplitude sweep, temperature sweep tests, and oscillatory time sweep are mostly conducted. While creep-recovery and relaxation tests are performed under transient mode. Unlike rotational techniques, oscillatory tests fundamentally evaluate the structural behavior of the mud at a micro-level. In such cases, the mud's response time can be investigated alongside the strain strength of the mud's inherent structures [3].

Generally, the rheometer comes with different geometries, a dual gap cylinder, smooth and rough parallel plate. They also have varying bob types (vane-type and bob-types) set at a specific gap depending on the geometry type. In addition, the volume of the sample to be used in rheometric evaluation also depends on the geometry. Rheometers are equipped with complex and accurate temperature regulated by a Peltier plate, which allows accurate heating and cooling rate adjustment. The type of rheometer geometry adversely affects hydrate-based rheology studies, especially in the case of a smooth Couette bob measuring geometry [6, 7]. Thus, authors often modify the smooth Couette bob geometry by making its surface rough to help form hydrates. However, the vane-type could be used if possible, at high-pressure conditions. Details of the viscoelastic measurement procedure are discussed below. Prior to running any rheometer, it is advisable to ensure that the setup is well-calibrated and that the sample is preconditioned appropriately. These will ensure good reading and data collection.

3.1.4.1 Steady Shear Test

Steady-shear tests are mostly conducted at a constant temperature to measure the shear stress and viscosity of the mud at a desired shear rate range. The specific shear rate range might vary depending on the kind of data needed. In the testing mode, the rheometer's temperature is set to its testing temperature and allowed to precondition the mud at a controlled shear property. Then, the experiment is started for the specified shear rates. At low shear rates (below 1 s^{-1}) the transient effect presence can be avoided or minimized by increasing the during for the systems generate data points. In this method, one advantage is to manipulate the variables in accordance with the desired analysis. For instance, testing can be conducted at a constant shear rate, shear stress, or viscosity, while varying other parameters [3].

3.1.4.2 Oscillatory Amplitude Sweep Tests

The amplitude test is the first test in the oscillatory test mode. It is conducted initially to evaluate the linear viscoelastic range. This is because the preceding oscillatory test would be conducted within the linear viscoelastic range. The viscoelastic range is the strain range at with the fluid exhibits a stable elastic and viscous behavior (G' and G''). The linear viscoelastic range in practice is the strain range at which the mud's internal structure can be restored or reversed to its normal state. The amplitude test is conducted at a constant frequency specified by the type of analysis to be conducted; however, a standard value of 10 rad/s is mostly used. At the constant frequency, the G' and G'' are measured by varying at a desired oscillatory strain range. The strain range is used to deform the mud until it assumes an irreversible deformation structure, known as the nonlinear viscoelastic response. The linear viscoelastic range represents the structural stability and strength of the mud's gelling characteristics. The mud's dynamic yield point can also be accurately determined from the amplitude test by determining the cross point of the flow point. The flow point is the intersection between the G' and G'' plot with oscillation strain. It must be noted that the amplitude test is a trial and error method [3].

3.1.4.3 Oscillatory Frequency Sweep Tests

The linear viscoelastic range determined in the amplitude test is held constant in the frequency test mode while ramping the frequency. The structural properties of the mud are testing in the viscoelastic range to understand the time-dependence and gel strength of the mud at rest [3].

3.1.4.4 Oscillatory Time Sweep Tests

This test is conducted to study the drilling mud's internal and gelling property response with time. In this mode, the viscoelastic properties of the are studied with time to understand the gelling speed and time of the drilling mud. Also, the development of the mud's internal structure and its time dependant dispersion settling ability can be studied. To perform the time sweep experiments, the frequency, temperature, and amplitude are kept constant with varying times. The amplitude and frequency are selected in the viscoelastic region from the amplitude and frequency sweep tests. The mud is initially pre-sheared, after which the time sweep is continued immediately after the pre-shearing. Just like the gelling time studies in the viscometer analysis, the time sweep test best provides accurate measurement of drilling mud's gel structure. In practice, it is used to describe the behavior of the mud to suspend cuttings and prevent barite sag when drilling is paused. The time taken for the mud to form a gel structure is evaluated in this test. Generally, a mud system that forms a gel structure quickly is desirable during drilling operations [3].

3.1.4.5 Oscillatory Temperature Sweep Tests

In this experiment, the frequency and amplitude are made constant while the mud is viscoelastic properties are measured with varying temperatures. In this study, the effect of temperature or the thermal stability of the gel structure of the mud is evaluated. In practice, the test provides relevant information on the thermally changes that could occur in the drilling mud while being flowed along the wellbore. Properties such as the ability of the mud to keep cutting in suspension can be monitored with temperature. Also, the barite sagging effect with temperature can be observed. This study can also be conducted in the steady-state sweep to monitor the dynamic viscosities response to temperature changes. The freezing of depression temperature of drilling muds can also be determined using this test, especially for drilling arctic and permafrost, and deep offshore regions. The freezing point of drilling muds should be lower than the wellbore temperature to prevent freezing or ice plug formation. However, this test can also be used to study hydrate and wax formation behavior in drilling muds [3].

3.1.4.6 Hydrate Tests (Rheometer Method)

Hydrate tests are conducted by first using the desired geometry [5]. Most of the rheometer geometries do not provide suitable stirring and flowability for hydrate formation. By far the vane type has proven to be best though it needs some further modifications [8]. Also, the bob geometer has been used but with a roughing surface to allow hydrate formation [7]. The temperature ramp and hold mode are used for the hydrate testing at a constant shear rate or amplitude and frequency. In this mode, the viscosity and/or viscoelastic properties are measured with first temperature reductions (temperature ramp) to the predetermined hydrate formation temperature (lowest wellbore temperature). When the temperature reaches the desired experimental temperature, the time sweep method is used to continuously measure the viscosity and/or viscoelastic properties changes of the mud with time every second. The rheometer provides a wide range of cooling and heating rate that could be adjusted to the cooling and heating rate of the drilling mud along the wellbore. Thus, allows an effective way to study hydrate formation and dissociation behavior in drilling muds. In order to detect the onset of hydrate formation, a sharp temperature in the mud system could be observed, however, this might not be very evident in most rheometers due to their effective temperature regulating technique. In such cases, a sharp increase in viscosity and/or viscoelastic properties is an indication of

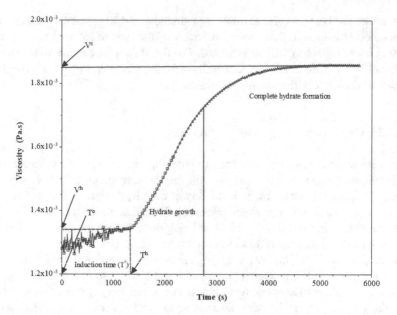

Fig. 3.1 Typical viscosity versus time profile during THF hydrate formation in drilling mud

the onset of hydrate formation. The hydrate formation is considered complete when the system viscosity becomes constant after a period of time (see Fig. 3.1). The hydrate test in drilling muds is conducted in high-pressure autoclave setups with a closed system. For ambient pressure hydrate formers like THF, the closed systems would help to prevent the evaporation of THF in the rheometer during testing. The time taken for hydrate to form and the maximum increase in the viscosity and/or viscoelastic properties of the mud are used to evaluate or quantify the amount of the hydrate formed.

The time taken for hydrate to form (induction or nucleation time), T^i is estimated from Fig. 3.1 by subtracting the time taken for hydrate to form (the time where the sharp temperature or viscosity increase was observed), T^h from the time the experiment started, T^o. V^m is the total amount of viscosity increased, which can be used to estimate the rate of the hydrate formation or viscosity increase with time. V^m is the subtraction of the V^c, maximum viscosity achieved during hydrate formation from V^h, the induction time viscosity [5].

3.2 Hydrate Testing Methods

The hydrate management properties of drilling mud are tested based on their desired performances as discussed in Chap. 2. In testing the effect of drilling mud on hydrates phase behavior, formation, and dissociation kinetics, different apparatus is mostly used. Some of the apparatus' are custom fabricated or purchased as ready to use.

3.2.1 Hydrate Testing Apparatuses

3.2.1.1 Apparatuses Differential Scanning Calorimetry (DSc)

The Differential Scanning Calorimetry (DSc) is one of the techniques that can be used to test the hydrate properties of drilling. The use of DSc for hydrate testing in drilling mud systems was introduced by Herzhaft and Dalmazzone [9] in 2001. The hydrate phase behavior of mud is most accurately measured using the DSc technique. DSc generally uses thermal characterization and heat exchange the mud sample and a reference against time or temperature to determine the hydrate phase boundary conditions. All DSc apparatuses consist of a sample and the reference in identical crucibles. The sample and references are designed such that they could be opened or closed to the atmosphere depending on the type of analysis of studies needed to be conducted. The DSc has a furnace that is thermally regulated and contains the two crucibles. However, different DScs might have different thermal exchange behavior, but thermocouples are used to measure the differences between sample and reference in the crucibles in simple DScs. The DSc is calibrated based on the heat exchange rate from the measured temperature difference between the sample and reference. However, heat exchanges in complex and modern DSc are measured by the Calvet principle. There are also temperatures sensors connected to the crucible to measure the heat flow exchanged between the furnace and the crucibles. Attached to most DSc apparatuses are pressure cells which could either be helium, nitrogen, methane, etc. A pressure controller is used to regulate the pressure in the cell during experimentation. In general, the samples holder capacity of DSc ranges from 20 to 50 μg. In some cases, the sample size serves as an advantage, while in drilling mud application it's not very representative of the bulk drilling mud systems and thus might lead to inaccurate formation measurement. The use of DSc in hydrate measurement, especially kinetics is characterized by wide errors, thus it's advisable to repeat the experiment more than 10 times.

3.2.1.2 Hydrate Pressure Cells

High-pressure hydrate cells are mostly used to test the hydrate properties of drilling muds. They are generally employed to study the hydrate formation and dissociation kinetics and phase behavior in drilling muds [10, 11]. Unlike the DSc, high-pressure hydrate cells are mostly used for hydrate studied and they are mostly customary fabricated by various research groups. Their fabrication design is based on the specific task need to measure with respect to the specific hydrate testing method. Depending on the type of analysis need, cameras and other types of analyzers are attached to the equipment for good results and visualization. The cells are mostly calibrated because they are custom fabricated. The phase behavior study is used for calibration, with the results compared with thermodynamic data predicted from CSMGem or PVTSim [12, 13].

Generally, the high-pressure hydrate cell employed to the hydrate in drilling mud systems is made of a stainless steel cylindrical or hemispherical reactor that is defined to a specific volume or capacity. However, the cell's capacities are large enough to allow hydrate formation, observation, and reduced the stochastic effect of hydrate nucleation. Pressure and temperature sensors are attached to the hydrate pressure cells for pressure and temperature measurement during experimentation. Similar to the DSc setup, a gas cylinder is connected to the setup which is controlled by a pressure regulator for safe pressurization. To control the temperature of the hydrate reactor, the cell is placed in a water bath system, whose temperature and is regulated by a thermostatic bath. For some reactors, agitation is provided via installing an adjustable magnetic stirrer with a wide range of speeds. When the apparatus is used for hydrate dissociation studies, a pump is attached to the cell for drilling mud injection purposes. The pumps can be manipulated to varying injection rates as desired. A data acquisition system is connected to the setup to record the changes in the system's parameters with time. However, other research attaches some analyzers to the cell for their preferred properties analysis. NMR and MRI are sometimes used to monitor the hydrate structures formed in the drilling mud systems.

3.2.2   Hydrate Testing Apparatuses

3.2.2.1   Hydrate Phase Behavior-DSc Technique

As stated earlier, the variation between the heat flow of the drilling mud sample and reference with respect to temperature phase changes is utilized to measure the hydrate dissociation temperature in drilling mud systems. The isothermal method is mostly used to measure the hydrate phase boundary points [9, 14–17]. Initially, the drilling mud is placed in the sample holder while the reference cell is empty. After that, the cells (sample and reference) are placed in the calorimetric block for purging to be done. Some amount of gas is initially used to purge the systems to remove any air present in the cell. The gas understudy is then pressurized into the cells to the desired experimental pressure. To determine the hydrate dissociation offset temperature, the systems in first cooling to a very low temperature at a specific cooling rate. The cooling rate is mostly fast, to allow hydrate formation or ice formation. The system is allowed to stay at the predetermined temperature for a long time to ensure hydrates formation. In the kinetic study, the system is cooling to the experimental temperature and not any arbitrary temperatures as conducted in the phase behavior studies. The systems are then gradually heated at a very slow heating rate to allow the hydrate to dissociate. Another significance of the slow heating rate is to allow a uniform hydrate dissociation as there is no stirring mechanism in any form. After heating the thermograms are recorded and analyzed to evaluate the hydrate dissociation temperature. However, hydrate quantitative analysis cannot be conducted in the DSc method, which limits its application in hydrate testing.

3.2.2.2 Hydrate Phase Behavior—Isochoric Pressure Search Technique

The isochoric pressure search or temperature search (T-cycle) method is used to determine the hydrate formation temperature in high-pressure hydrate cells [11, 18–28]. This constant volume method ensured a fixed cell with a known amount of gas or hydrate former. In this mode, the cell is first purged or vacuumed after loading the mud sample in the reactor. In this cell, the sample sizes a very large compared with the DSc. The cell is pressurized with the gas and allowed to stabilize at the experimental conditions (mostly pressure and temperature close to the expected hydrate equilibrium temperature). The systems are mixed by turning the mixer on. The hydrate formation test is initiated by cooling the system temperature to a temperature low enough to provide the required driving force for hydrate formation to occur. After the hydrates are formed (which can be observed visually or via the cell temperature and pressure changes), the system is slow heated stepwise at and very slow temperature stepwise up to the initial experimental temperature. The hydrate dissociation temperature is taken as the intersection between the heating and cooling curve.

3.2.2.3 Kinetics of Hydrate Formation

The hydrate formation kinetic in drilling mud is tested by using the isochoric constant cooling method [22, 26, 29–39]. After cleaning, purging, and setting up the cell in the systems, the desired gas is pressurized in the cell with the stirrer turned on at the initial experimental conditions. Similar to the phase behavior study, the temperature of the system is cooled to the experimental temperature at a constant cooling rate. The pressure and temperature in the system are monitored continuously by the data acquisition system. Hydrate formation is then detected by a sharp increase in the system temperature or a sharp decrease in pressure. For some setups with visual capabilities, the hydrate formation can be detected visually. When a constant pressure and temperature is achieved in the system after 5 h or more, the hydrate formation process is considered complete. The data is then extracted and analyzed.

3.2.2.4 Kinetics of Hydrate Formation

Mostly different methods are used by various authors to study the dissociation kinetics of hydrate in drilling mud. Generally, this is test is conducted by injecting a precooled drilling mud system into the cell after hydrate formation with a constant flux pump [29, 30, 40, 41]. The drilling mud is precooled temperature to the same as the experimental temperature. The injection of the drilling mud into the reactor is performed to test the ability of the mud to minimize or prevent hydrate dissociation in the wellbore. While the injecting is taken place, the changes in the reactor are recorded to study the dissociation rate of the hydrate. The hydrate dissociation is noticed by an increase in the reactor pressure with time. In other methods, the temperature of the systems is increased to high temperature at a controlled rate to simulate increasing downhole temperature caused by heat transfer between the mud and hydrate rock. Such studies only study the effect of the temperature exchange between the mud and water. It does not account for the chemical dissociation effect, but the thermal dissociation effect of the mud. The heating simulations also help to mimic the effect of heat transfer from drill assemblies and the wellbore. When the temperature of the system is increased, the hydrates are dissociated by observing an increase in the system's pressure. When the pressure in the system is constant the dissociation process is considered complete.

However, the effect of the drilling mud filtrate on the dissociation of the hydrate is less tested. The available testing in such regard is based on testing the effect of the mud on the phase behavior of the hydrates. However, such testing thus does not quantify the exact amount of hydrate released from the sediment. The rate of gas released for the sediment cannot be evaluated in such a mode. Thus, proposing an accurate technique to measure such chemical hydrate dissociation effect in hydrate sediments is useful for fully understand the effect of drilling mud methane hydrate sediments.

References

1. Paul Scott, Broussard P, Bland R, Growcock F, Freeman M (2015) Drilling fluids. In: IADC drill. man., 12th edn, p 66
2. API-RP13B-2 (2014) Recommended practice for field testing oil-based drilling fluids. In: 5th edn, p 141
3. Bui B (2012) Viscoelastic properties of drilling fluids. Annu Trans Nord Rheol Soc 20:33–47
4. Werner B, Myrseth V, Saasen A (2017) Viscoelastic properties of drilling fluids and their influence on cuttings transport. J Pet Sci Eng 156:845–851
5. Bavoh CB, Ofei TN, Lal B (2020) Investigating the potential cuttings transport behavior of ionic liquids in drilling mud in the presence of sII hydrates. Energy Fuels 34:2903–2915
6. Karanjkar PU, Ahuja A, Zylyftari G, Lee JW, Morris JF (2016) Rheology of cyclopentane hydrate slurry in a model oil-continuous emulsion. Rheol Acta 55:235–243
7. Pandey G, Linga P, Sangwai JS, Pandey G, Linga P, Sangwai JS (2017) High pressure rheology of gas hydrate formed from multiphase systems using modified Couette rheometer High pressure rheology of gas hydrate formed from multiphase systems using modified Couette rheometer. Rev Sci Instrum 88:025102
8. Jorge P, Karanjkar PU, Lee JW, Morris JF (2010) Rheology of hydrate forming emulsions. Langmuir 26:11699–11704
9. Herzhaft B, Dalmazzone C (2001) Gas hydrate formation in drilling mud characterized with DSC technique. In: SPE annu. tech. conf. exhib. New Orleans, Louisiana, pp 575–584
10. Partoon B, Sabil KM, Lau KK, Lal B, Nasrifar K (2018) Production of gas hydrate in a semi-batch spray reactor process as a means for separation of carbon dioxide from methane. Chem Eng Res Des 138:168–175

11. Bavoh CB, Lal B, Keong LK (2020) Introduction to gas hydrates. In: Chem. addit. gas hydrates. Green energy and technology. Springer, Cham, pp 1–20

12. Bavoh CB, Khan MS, Ting VJ, Lal B, Ofei TN, Ben-Awuah J, Ayoub M, Shariff ABM (2018) The effect of acidic gases and thermodynamic inhibitors on the hydrates phase boundary of synthetic Malaysia natural gas. IOP Conf Ser Mater Sci Eng 458:012016

13. Broni-Bediako E, Amorin R, Bavoh CB (2017) Gas hydrate formation phase boundary behaviour of synthetic natural gas system of the Keta basin of Ghana. Open Pet Eng J 10:64–72

14. Xiao C, Wibisono N, Adidharma H (2010) Dialkylimidazolium halide ionic liquids as dual function inhibitors for methane hydrate. Chem Eng Sci 65:3080–3087

15. Sabil KM, Nashed O, Lal B, Ismail L, Japper-Jaafar A, Nashed O, Lal B, Ismail L (2015) Experimental investigation on the dissociation conditions of methane hydrate in the presence of imidazolium-based ionic liquids. J Chem Thermodyn 84:7–13

16. Nashed O, Dadebayev D, Khan MS, Bavoh CB, Lal B, Shariff AM (2018) Experimental and modelling studies on thermodynamic methane hydrate inhibition in the presence of ionic liquids. J Mol Liq 249:886–891

17. Nashed O, Sabil KM, Lal B, Ismail L, Jaafar AJ (2014) Study of 1-(2-Hydroxyethyle) 3-methylimidazolium Halide as thermodynamic inhibitors. Appl Mech Mater 625:337–340

18. Bavoh CB, Khan MS, Lal B, Bt Abdul Ghaniri NI, Sabil KM (2018) New methane hydrate phase boundary data in the presence of aqueous amino acids. Fluid Phase Equilib 478:129–133

19. Khan MS, Bavoh CB, Partoon B, Lal B, Bustam MA, Shariff AM (2017) Thermodynamic effect of ammonium based ionic liquids on CO_2 hydrates phase boundary. J Mol Liq 238:533–539

20. Khan MS, Lal B, Keong LK, Ahmed I (2019) Tetramethyl ammonium chloride as dual functional inhibitor for methane and carbon dioxide hydrates. Fuel 236:251–263

21. Khan MS, Partoon B, Bavoh CB, Lal B, Mellon NB (2017) Influence of tetramethylammonium hydroxide on methane and carbon dioxide gas hydrate phase equilibrium conditions. Fluid Phase Equilib 440:1–8

22. Bavoh CB, Lal B, Khan MS, Osei H, Ayuob M (2018) Combined inhibition effect of 1-ethyl-3-methy-limidazolium chloride + glycine on methane hydrate. J Phys Conf Ser 1123:012060

23. Yuha YBM, Bavoh CB, Lal B, Broni-Bediako E (2020) Methane hydrate phase behaviour in EMIM-Cl water based mud (WBM): an experimental and modelling study. S Afr J Chem Eng 34:47–56

24. Mannar N, Bavoh CB, Baharudin AH, Lal B, Mellon NB (2017) Thermophysical properties of aqueous lysine and its inhibition influence on methane and carbon dioxide hydrate phase boundary condition. Fluid Phase Equilib 454:57–63

25. Bavoh CB, Partoon B, Lal B, Keong LK (2016) Methane hydrate-liquid-vapour-equilibrium phase condition measurements in the presence of natural amino acids. J Nat Gas Sci Eng 37:425–434

26. Bavoh CB, Yuha YB, Tay WH, Ofei TN, Lal B, Mukhtar H (2019) Experimental and modelling of the impact of quaternary ammonium salts/ionic liquid on the rheological and hydrate inhibition properties of Xanthan gum water-based muds for drilling gas hydrate-bearing rocks. J Pet Sci Eng 183:106468

27. Bavoh CB, Partoon B, Lal B, Gonfa G, Foo Khor S, Sharif AM (2017) Inhibition effect of amino acids on carbon dioxide hydrate. Chem Eng Sci 171:331–339

28. Bavoh CB, Nashed O, Khan MS, Partoon B, Lal B, Sharif AM (2018) The impact of amino acids on methane hydrate phase boundary and formation kinetics. J Chem Thermodyn 117:48–53

29. Srungavarapu M, Patidar KK, Pathak AK, Mandal A (2018) Performance studies of water-based drilling fluid for drilling through hydrate bearing sediments. Appl Clay Sci 152:211–220

30. Zhao X, Qiu Z, Zhao C, Xu J, Zhang Y (2019) Inhibitory effect of water-based drilling fluid on methane hydrate dissociation. Chem Eng Sci 199:113–122

31. Saikia T, Mahto V (2018) Experimental investigations and optimizations of rheological behavior of drilling fluids using RSM and CCD for gas hydrate-bearing formation. Arab J Sci Eng 43:1–14

32. Ning F, Zhang L, Tu Y, Jiang G, Shi M (2010) Gas-hydrate formation, agglomeration and inhibition in oil-based drilling fluids for deep-water drilling. J Nat Gas Chem 19:234–240

33. Saikia T, Mahto V (2016) Evaluation of 1-decyl-3-methylimidazolium tetrafluoroborate as clathrate hydrate crystal inhibitor in drilling fluid. J Nat Gas Sci Eng 36:906–915
34. Jiang G, Liu T, Ning F, Tu Y, Zhang L, Yu Y, Kuang L (2011) Polyethylene glycol drilling fluid for drilling in marine gas hydrates-bearing sediments: an experimental study. Energies 4:140–150
35. Saikia T, Mahto V (2016) Experimental investigations of clathrate hydrate inhibition in water based drilling fluid using green inhibitor. J Pet Sci Eng 147:647–653
36. Bavoh CB, Lal B, Ben-Awuah J, Khan MS, Ofori-Sarpong G (2019) Kinetics of mixed amino acid and ionic liquid on CO_2 hydrate formation. In: IOP Conf. Ser. Mater. Sci. Eng. p 012073
37. Bavoh CB, Ntow T, Lal B, Sharif AM, Shahpin MHBA, Sundramoorthy JD (2019) Assessing the impact of an ionic liquid on NaCl/KCl/polymer water-based mud (WBM) for drilling gas hydrate-bearing sediments. J Mol Liq 294:111643
38. Khan MS, Cornelius BB, Lal B, Bustam MA (2018) Kinetic assessment of tetramethyl ammonium hydroxide (Ionic Liquid) for carbon dioxide, methane and binary mix gas hydrates. In: Rahman MM (ed) Recent Adv. Ion. Liq. IntechOpen, London, UK, pp 159–179
39. Bavoh CB, Lal B, Keong LK, Jasamai MB, Idress MB (2016) Synergic kinetic inhibition effect of EMIM-Cl + PVP on CO_2 hydrate formation. In: Procedia Eng. pp 1232–1238
40. Kawamura T, Yamamoto Y, Yoon JH, Haneda H, Ohga K, Higuchi K (2002) Dissociation behavior of methane-ethane mixed gas hydrate in drilling mud fluid. Proc Int Offshore Polar Eng Conf 12:439–442
41. Fereidounpour A, Vatani A (2014) An investigation of interaction of drilling fluids with gas hydrates in drilling hydrate bearing sediments. J Nat Gas Sci Eng 20:422–427

Chapter 4
Kinetics Behaviour of Hydrates Drilling Muds

In this section, the kinetics data on managing hydrate formation and dissociation in drilling mud while drilling hydrate sediment are reviewed and discussed. The data provided in this section provides fundamental information on the effect of drilling mud additives on the kinetics of hydrates formation and dissociation in the presence of hydrate different hydrate formers.

4.1 Kinetics of Hydrate Formation and Dissociation in Drilling Mud

Table 4.1 presents the reported data of the hydrate formations and dissociations kinetics of different drilling mud additives in varying gas systems. The presences of oil in oil-based muds delays hydrate formation in water to an order of 360%. However, a sharp and high amount of hydrates are formed in oil systems on the onset of hydrate are formed in oil systems [1]. The rapid rate of gas-hydrates formation in the oil system is due to the high solubility of the gas in the oil phase and the large surface area of the dispersed water phase, Ebeltoft et al. [2] further confirms this phenomenon. Furthermore, hydrates formed in oil-based drilling fluids induce phase separation because hydrates start forming at the interface between the dispersed water droplets and the continuous oil phase. This will induce a dramatic change in the mechanical properties of the interfacial film, and hence, there will be no barrier towards coalescence and phase separation [2]. However, diesel has a negligible effect on hydrate formation rate [3]. The presence of $CaCl_2$ brine could reduce the amount of hydrates formed in oil–water and mud systems. Lignosulfonate and lignite slightly promote hydrate nucleation time in water base mud. This is because of their dispersed nature in WBMs, thus making it easier to form hydrates in water [3]. Though they could reduce hydrate formation rate and gas uptake. However, the presence of salt reduces the hydrate formation rate. NaCl, glycerine, and propylene glycol slow down

B. Lal et al., *Hydrate Control in Drilling Mud*,
SpringerBriefs in Petroleum Geoscience & Engineering,
https://doi.org/10.1007/978-3-030-94130-7_4

Table 4.1 Reported kinetics data on drilling mud in presence of hydrate inhibitors

Author	Gas	System	Induction time		Uptake		Rate of formation/dissociation	
			Exp	RIE	Exp	RIE	Exp	RIE
Schofield et al. [6]	Natural gas[1]	DI	NA	NA	NA	NA	0.042^c	Base
		0.2^aLecithin	NA	NA	NA	NA	0.026^c	38.1
		0.2^aLecithin + 0.1^aGlycerol	NA	NA	NA	NA	0.026^c	38.1
		0.2^aLecithin + 0.02^aPVP	NA	NA	NA	NA	0.059^c	−40.48
		0.2^aLeicithin + 0.02^aPVP + 0.1^aConDet + 0.02^aDefoamer	NA	NA	NA	NA	0.018^c	57.14
		0.02^aLeicithin + 0.1^aConDet + 0.02Defoamer	NA	NA	NA	NA	0.048^c	−14.29
Zhao et al. [9]	CH$_4$	BM-1	$6.1^{(DT)}$	NA	NA	NA	$0.379^{(DT)}$	
		BM-1 + 0.1^aPVP	$9.2^{(DT)}$	50.9	NA	NA	$0.257^{(DT)}$	47.47
		BM-1 + 0.25^aPVP	$9.8^{(DT)}$	60.7	NA	NA	$0.243^{(DT)}$	55.97
		BM-1 + 0.5^aPVP	$10.4^{(DT)}$	70.5	NA	NA	$0.221^{(DT)}$	71.49
		BM-1 + 0.75^aPVP	$10.4^{(DT)}$	70.5	NA	NA	$0.223^{(DT)}$	69.96
		BM-1 + 1^aPVP	$10.9^{(DT)}$	78.7	NA	NA	$0.214^{(DT)}$	77.10
		BM-1 + 5^aEG	$6.5^{(DT)}$	6.56	NA	NA	$0.351^{(DT)}$	7.98
		BM-1 + 10^aEG	$7.6^{(DT)}$	24.6	NA	NA	$0.311^{(DT)}$	21.86
		BM-1 + 15^aEG	$6.1^{(DT)}$	0.00	NA	NA	$0.382^{(DT)}$	−0.79
		BM-1 + 20^aEG	$5.8^{(DT)}$	−4.92	NA	NA	$0.407^{(DT)}$	−6.88
		BM-1 + 30^aEG	$5.3^{(DT)}$	−13.1	NA	NA	$0.443^{(DT)}$	−14.45
		BM-1 + 0.1^aLecithin	$7.7^{(DT)}$	26.2	NA	NA	$0.297^{(DT)}$	27.61

(continued)

Table 4.1 (continued)

Author	Gas	System	Induction time		Uptake		Rate of formation/dissociation	
			Exp	RIE	Exp	RIE	Exp	RIE
		BM-1 + 0.25[a]Lecithin	8.5[(DT)]	39.3	NA	NA	0.274[(DT)]	38.32
		BM-1 + 0.5[a]Lecithin	9.4[(DT)]	54.1	NA	NA	0.238[(DT)]	59.24
		BM-1 + 0.75[a]Lecithin	9.8[(DT)]	60.7	NA	NA	0.229[(DT)]	65.50
		BM-1 + 1[a]Lecithin	10.4[(DT)]	70.4	NA	NA	0.216[(DT)]	75.46
		BM-1	8.0[(DT)]	31.15	NA	NA	0.31[(DT)]	22.26
		BM-1 + 0.1[a]PVP	10.7[(DT)]	33.75	NA	NA	0.225[(DT)]	37.78
		BM-1 + 0.5[a]Lecithin	13.2[(DT)]	65.00	NA	NA	0.171[(DT)]	81.29
		BM-1 + 0.1[a]PVP + 0.5[a]Lecithin	14.8[(DT)]	85.00	NA	NA	0.154[(DT)]	101.30
		BM-1 + 0.5[a]PVP + 0.5[a]Lecithin	16.7[(DT)]	108.8	NA	NA	0.133[(DT)]	133.08
		BM-1 + 10[a]EG	10.6[(DT)]	32.50	NA	NA	0.224[(DT)]	38.39
Zhao et al. [10]	Natural gas[2]	BM-2	61[R1]	Base	NA	NA	3.61	Base
		BM-2 + 10[a]NaCl	>900[R1]	1539	NA	NA	1.1	69.53
			>900[R2]	1539	NA	NA	1.07	70.36
		BM-2 + 10[a]Glycol	170[R1]	178.7	NA	NA	2.59	28.25
			229[R2]	275.4	NA	NA	2.16	40.17
			185[R3]	203.3	NA	NA	2.4	33.52
		BM-2 + 10[a]Glycerol	112[R1]	83.61	NA	NA	2.85	21.05
			141[R2]	131.2	NA	NA	2.55	29.36
Saikia and Mahto [11]	THF	BM-3	5	Base	NA	NA	NA	NA

(continued)

Table 4.1 (continued)

Author	Gas	System	Induction time		Uptake		Rate of formation/dissociation	
			Exp	RIE	Exp	RIE	Exp	RIE
		BM-3 + 0.1aPVP	11	120	NA	NA	NA	NA
		BM-3 + 0.5aPVP	590	11,700	NA	NA	NA	NA
		BM + 1aPVP	1440	28,700	NA	NA	NA	NA
			46	820	NA	NA	NA	NA
			700	13,900	NA	NA	NA	NA
			1440	28,700	NA	NA	NA	NA
			1440	28,700	NA	NA	NA	NA
			1440	28,700	NA	NA	NA	NA
			1440	28,700	NA	NA	NA	NA
Saikia and Mahto [11]	THF	BM-3	5	Base	NA	NA	NA	NA
		BM-3 + 0.1aPVP	12	140	NA	NA	NA	NA
		BM-3 + 0.5aPVP	600	11,900	NA	NA	NA	NA
		BM + 1aPVP	1440	28,700	NA	NA	NA	NA
		BM-3 + 0.1aDMIMBF4	59	1080	NA	NA	NA	NA
		BM-3 + 0.5aDMIMBF4	1440	28,700	NA	NA	NA	NA
		BM-3 + 1aDMIMBF4	1440	28,700	NA	NA	NA	NA
		BM-3 + 0.1aDMIMBF4 + 0.1aPVP	85	1600	NA	NA	NA	NA
		BM-3 + 0.5aDMIMBF4 + 0.1aPVP	1440	28,700	NA	NA	NA	NA

(continued)

Table 4.1 (continued)

Author	Gas	System	Induction time		Uptake		Rate of formation/dissociation	
			Exp	RIE	Exp	RIE	Exp	RIE
		BM-3 + 1aDMIMBF4 + 0.1aPVP	1440	28,700	NA	NA	NA	NA
Saikia and Mahto [12]	THF	BM-4	46	820	NA	NA	NA	NA
		BM-3 + 0.1aPVP	700	13,900	NA	NA	NA	NA
		BM-3 + 0.5aPVP	1440	28,700	NA	NA	NA	NA
		BM-4 + 1aPVP	1440	28,700	NA	NA	NA	NA
		BM-4 + 0.1aLipase	1440	28,700	NA	NA	NA	NA
		BM-4 + 0.5aLipase	1440	28,700	NA	NA	NA	NA
		BM-4 + 1aLipase	5	Base	NA	NA	NA	NA
		BM-4 + 0.1alipase + 0.1aPVP	12	140	NA	NA	NA	NA
		BM-4 + 0.5alipase + 0.1aPVP	600	11,900	NA	NA	NA	NA
		BM-4 + 1alipase + 0.1aPVP	1440	28,700	NA	NA	NA	NA
Saikia and Mahto [11]	THF	BM-3	59	1080	NA	NA	NA	NA
		BM-3 + 0.1aPVP	1440	28,700	NA	NA	NA	NA
		BM-3 + 0.5aPVP	1440	28,700	NA	NA	NA	NA
		BM + 1aPVP	85	1600	NA	NA	NA	NA
		BM-3 + 0.1aDMIMBF4	1440	28,700	NA	NA	NA	NA
		BM-3 + 0.5aDMIMBF4	1440	28,700	NA	NA	NA	NA

(continued)

Table 4.1 (continued)

Author	Gas	System	Induction time		Uptake		Rate of formation/dissociation	
			Exp	RIE	Exp	RIE	Exp	RIE
		BM-3 + 1aDMIMBF4	1440	28,700	NA	NA	NA	NA

Author	Gas	System	Induction time		Uptake		Rate of formation/dissociation	
			Exp	RIE	Exp	RIE	Exp	RIE
Lai and Dzialowski [3]	Natural gas[3]	BM-5	157	Base	NA	NA	NA	NA
		BM-5 + 30 ppb gel	77	−50.9	NA	NA	NA	NA
		BM-5 + 10 ppb salt	52	−66.8	NA	NA	NA	NA
		BM-5 + 50 ppb salt	77	−50.9	NA	NA	NA	NA
		BM-5 + 90 ppb salt	167	6.37	NA	NA	NA	NA
		BM-5 + Lignosulfonate	55	−64.9	NA	NA	NA	NA
		BM-5 + Lignite	52	−66.8	NA	NA	NA	NA
		BM-5 + 5ppbGel + Caustic)	257	63.69	NA	NA	NA	NA
		BM-5 + 20 ppb Gel)	182	15.92	NA	NA	NA	NA
		BM-5 + 2%Diesel	67	−57.3	NA	NA	NA	NA
		Polymer A	NHF	NHF	NA	NA	NA	NA
		Polymer B	NHF	NHF	NA	NA	NA	NA
		Polymer C	537	242.0	NA	NA	NA	NA
		Polymer D	NHF	NHF	NA	NA	NA	NA
		Polymer E	279	77.71	NA	NA	NA	NA
		Seawater gyp	62	−60.5	NA	NA	NA	NA

(continued)

Table 4.1 (continued)

Author	Gas	System	Induction time		Uptake		Rate of formation/dissociation	
			Exp	RIE	Exp	RIE	Exp	RIE
		Lime mud	50	−68.2	NA	NA	NA	NA
		BM-5 + 50 ppb glycerine	157	0.00	NA	NA	NA	NA
		BM-5 + 50 vol% glycerine	102	−35.0	NA	NA	NA	NA
		BM-5 + 100 ppb propylene glycol	207	31.85	NA	NA	NA	NA
		BM-5 + 50 vol% propylene glycol	866	451.6	NA	NA	NA	NA
Zhang et al. [9]	CH_4	BM-6	85	Base	NA	NA	1.03	Base
		BM-6 + 10% NaCl	70	−17.65	NA	NA	1.29	−25.24
		BM-6 + 15% NaCl	60	−29.41	NA	NA	1.38	−33.98
		BM-6 + 10% EG	80	−5.88	NA	NA	1.11	−7.77
		BM-6 + 15% EG	75	−11.76	NA	NA	1.18	−14.5
Mech and Sangwai [5]	CH_4	pure water (c = 7.5 MPa)	21.6	Base	NA	NA	85.8	Base
		pure water (c = 5.5 MPa)	11.7	Base	NA	NA	55.1	Base
		0.2 wt% PEG-200(c = 7.5 MPa)	23.4	8.3	NA	NA	56.0	34.6
		0.2 wt% PEG-200(c = 5.5 MPa)	16.8	43.6	NA	NA	35.5	35.4
		0.4 wt% PEG-200(c = 7.5 MPa)	35.7	65.3	NA	NA	52.9	38.3
		0.4 wt% PEG-200(c = 5.5 MPa)	28.2	114	NA	NA	34.7	36.9

(continued)

Table 4.1 (continued)

Author	Gas	System	Induction time		Uptake		Rate of formation/dissociation	
			Exp	RIE	Exp	RIE	Exp	RIE
		0.2 wt% PEG-600(c = 7.5 MPa)	19.5	−9.7	NA	NA	48.8	43.1
		0.2 wt% PEG-600(c = 5.5 MPa)	24.6	110	NA	NA	23.1	58
		0.4 wt% PEG-600(c = 7.5 MPa)	89.4	313.9	NA	NA	54.8	36.2
		0.4 wt% PEG-600(c = 5.5 MPa)	51.6	341	NA	NA	38.7	29.8
Dzialowski et al. [4]	CH_4	BM-7 (Gulf of Mexico)	540	Base	NA	NA	NA	NA
		BM-7 + KI	1815	236.1	NA	NA	NA	NA

(DT)Dissociation test; Base mud (BM); De-ionized water (DI); Methane (CH_4); 1-3Decyl-Methylimidazolium Tetrafluoroborate (DMIMBF$_4$); Experimental data (Exp.); Relative inhibition efficiency (RIE); Run (R); Not applicable (NA); ^{a}wt%; ^{b}mol/h; ^{c}gmol/h; ^{1}(94C1 + 4C$_2$ + 2C$_3$); ^{2}(80.42C$_1$ + 10.35C$_2$ + 1.82CO$_2$ + 0.11N$_2$); BM-1 (4% bentonite + 0.25 wt% FA-367 + 0.5wt%PAC-LV + 2 wt% SMP + 5 wt%KCl + 10 wt%NaCl); BM-2 (4% bentonite + 0.1 wt% XG + 0.4 wt%PAC + 0.4 wt% starch + 4 wt%KCl + 0.3 wt%NaOH)

BM-5 (25 ppb Gel + 1 ppb Lignite + 2 ppb Lignosulfonate)

^{3}(87.55C$_1$ + 7.59C$_2$ + 3.1C$_3$ + 0.19i-C$_4$ + 0.79n-C$_4$ + 0.19i-C$_5$ + 0.19n-C$_6$ + 0.39N$_2$); BM-5 (25 ppb Gel + 1 ppb Lignite + 2 ppb Lignosulfonate); Polymer A (10 ppb Gel + 0.25 ppb Caustic + 1.5 ppb PHPA + 1 ppb Starch + 0.25 ppb PAC); Polymer B (10 ppb Gel + 0.25 ppb Caustic + 0.75 ppb PHPA + 1 ppb Starch + 0.25 ppb PAC, MW = 10 ppg); Polymer C (1.25 ppb PHPA + 100 ppb Salt + 1.0 ppb PAC + 1.5 ppb Caustic + 0.35 ppb Soda Ash + 0.15 ppb Bicarb. + 20 ppb Gel + 21 ppb Rev Dust); Polymer D (Sea Water + 1 0 ppb Prehydrated Gel + 0.25 Caustic + 2.0 ppb PHPA + 3.5 ppb Starch + 0.5 ppb PAC); Polymer E (Polymer Mud D + 74.8 ppb Salt); 10 ppb salt (+100 ppb Salt + 1 ppb Lignosulfonate); Lignite (35 ppb Gel + 2 ppb Lignite); Lignosulfonate (35 ppb Gel + 2 ppb Lignosulfonate, pH = 10); 90 ppb salt (90 ppb Salt + 1 ppb Lignosulfonate); 2%Diesel (5 ppb Gel + 2%Diesel + 0.3%Emulsifier); Sea water gyp (Sea Water + 25 ppb Prehydrated Gel + 4 ppb Lignite + 3 ppb Lignosulfonate + 8 ppb Gypsum + 0.5 ppb CMC); Lime mud (20 ppb Gel + 3 ppb Caustic + 3 ppb Lignosulfonate + 2 ppb Lignite + 8 ppb Lime + 0.75 ppb CMC); No hydrate formed (NHF); BM-6 (4% bentonite + 0.3% ZCJ-1 + 0.1% ZCJ-2 + 0.5% JYJ-1 + 3% JYJ-2 + 3% SD-102) 50 vol%glycerine (50/50 by Volume of glycerine and Base Mud); 50 vol% propylene glycol (50/50 by Volume of Propylene Glycol and Base Mud); BM-7 (Gulf of Mexico); Kinetic inhibitor (KI); ^{d}mmol/mol the reaction rate at high concentrations. While caustic and gel have very less effect on gas hydrate formation reaction rates.

Dzialowski et al. [4] tested the effect of a known kinetic inhibitor on the natural gas hydrate system. The inhibitor exhibited a dual functional gas hydrate inhibitory property which is very suitable for deep-sea drilling operations where conventional THI are ineffective at higher concentrations. The inhibitor could reduce the hydrate formation temperature by 8 K and delay the hydrate nucleation time for 10 h at very high subcooling conditions. At moderate condition, hydrates formation are completely prevented. Interestingly, the inhibitor was compatible with conventional drilling fluid formulations that contain solids, salts, and polymers. Typically, such drilling fluid could reduce the concentrations of inhibitors, avoid the use of salts and glycols in some cases due to their dual-functional behaviour [4]. However, such inhibitors are not suitable for drilling hydrate sediments because of the thermodynamic inhibition effect, which could destabilize the hydrate sediments and lead to wellbore instability.

The use of PEG has proved as a good gas hydrate inhibitor for water systems at high concentrations. Especially, PEG with high molecular weight (PEG 600) is very effective in hydrate inhibition at lower concentrations. They are able to delay the hydrate nucleation time and reduce the amount of gas uptake into hydrates. This shows their possible potential as a gas drilling mud additive for drill gas hydrate sediments since they could reduce the possibility of the reformation of gas hydrate in the wellbore [5]. However, it is recommended to test their compatibility with standard drilling mud additives. As the molecular weight of the PEG increases, the viscous nature is enhanced which causes an increasing gas hydrate inhibition. However, they perform very well at low subcooling conditions, thus their application in high subcooling environments could be ineffective. The use of high molecular weight polymers might greatly alter the rheological properties of the mud.

Lecithin is known to lower the hydrate formation rate in water [6], however, it has negligible synergy with glycerol. On the other hand, the synergy of lecithin and PVP enhances hydrate formation by about 43% compared with water system. However, when other drilling fluid constituents like Con-Det and Bara Defoamer are added lecithin and PVP, the hydrate formation rate is greatly reduced. On the contrary, the presence of lecithin in drilling mud promotes hydrate formation and reduces gas hydrate dissociation rate. Lecithin can reduce hydrate dissociation rate via its surface adsorption mechanism. Their surface nature makes them form potential nucleation sites and thus promote hydrate formation rate. This promotes their use in drilling gas hydrate sediments or permafrost areas. Lecithin drilling mud systems have successfully been used in drilling the permafrost (the Cascade development in the Alaskan Arctic) and has proven to work effectively and save drilling cost as well. However, their hydrate-promoting nature might be a great treat when used in areas where there is a lot of gas in the wellbore and during shut-in operations. During shut-in or the presence of excess gas hydrate reformation, wellbore plug might occur easily since lecithin kinetically promotes hydrate formation. MMT promotes hydrate formation by providing sites for hydrate nucleation. This does not only occur by surface sites but there is evidence of internal microstructure formation of methane hydrate formation sites in the MMT layers [7]. It normally consists of sI hydrate structure with 3 water hydrate layers. This means that MMT does not significantly affect the hydrate structure, hence might negligible/slightly affect the hydrate equilibrium curve [7]. In addition, hydrate particles easily adsorb to the surface of clay particles dispersed in the seawater, which promotes hydrate growth and aggregation. With the increment of clay, these "promotion effect" become stronger. However, the presence of clay act as a thermodynamic promoter.

The effect of drilling mud on hydrate dissociation is critically dependent on the mud filtrate volumes and rate. A huge fluid loss property causes the hydrate dissociation rate. Suitable mud for hydrate drilling must have a good rheological property and low mud filtration while all of them were thermally stable [8]. The dissociation of hydrate sediments is governed by heat transfer phenomena, which is related to the thermal conductivity of the mud. The unsettling nature of hydrate and response to hydrate dissociation by the drilling mud may result in hydrate reformation due to the presence of gas and water in the hydrate stable condition along the annulus. This is caused by the pressure and temperature conditions in the annulus. This could lead to drilling challenges such as drill pipes stuck, plugging kill line and BOP, etc. Hence, the need for hydrate sediment drilling mud that exhibits kinetic hydrate inhibition properties, since the use of thermodynamic inhibitors might cause further dissociation of hydrates.

The presence of NaOH, KCl, and caustic soda increases the hydrate dissociation which consequently leads to wellbore cave and probability collapse [8, 9]. Inorganic salts promote hydrate dissociation because the addition of salts reduces water activity and the hydrate equilibrium curve shifts towards lower temperatures and higher pressures [9]. On the other hand, the use of polymers such as starch and PHPA reduces the mud hydrate dissociation rate. This is probably because they have good filtration management properties and act as a strong kinetic gas hydrate inhibitor. It is recommended that the use of PEG and EG at low concentrations (<5 wt%) to manage hydrate reformation in drilling facilities [8, 9]. On contrary, Zhao et al. [9] indicated that 10 wt% for EG solution significantly delays hydrate dissociation rate about 21% compared with water. However, this finding contradicts the general behavior of THIs on hydrate dissociation behavior, thus requires further investigations. However, at elevated concentrations, EG significantly increases hydrate dissociation. The use of commercial gas hydrate polymers such as PVP and lecithin does not only delay hydrate formation but PVP is able to delay hydrate dissociation rate at low concentrations. PVP and lecithin can reduce hydrate dissociation time and rate about 47–78% and 26–65%, respectively. compared with water. This means that PVP has dual kinetic inhibition functional behavior. i.e. they can kinetically delay hydrate formation and dissociation at the same time. This is very desirable for drilling methane hydrate sediments as it can maintain wellbore stability and prevent hydrate reformation as well. However, at higher or severe conditions PVP can rapidly form hydrate once its strength is exhausted.

As a typical KHI, PVP and lecithin molecules can adsorb onto the hydrate surface through hydrogen bonding. This forms a polymer film on the hydrate surface and fits the five-membered ring into the hydrate cage, hindering cavity completion and thus inhibiting hydrate growth. In the same way, during hydrate dissociation, PVP can be adsorbed onto the hydrate surface to form a film that can hinder the transfer of water and gas molecules from the hydrate bulk into the liquid phase, thereby slowing hydrate dissociation. In addition, as a type of high molecular weight polymer, the addition of PVP and lecithin significantly increases the liquid viscosity. High viscosity in turn increases the mass transfer resistance. Moreover, the addition of a small amount of PVP did not influence the thermal conductivity of the solution, indicating that PVP and lecithin do not inhibit hydrate dissociation by influencing heat transfer. However, the synergic impact of both PVP and lecithin at 0.5 wt% is good in hydrate dissociation management during drilling. Both PVP and lecithin showed promising compatibility with drilling mud additives at low concentrations (<0.5wt%). Kawamura et al. [13] also postulated that the presence of XANVIS water-based mud reduces hydrate dissociation with increasing concentrations. The decrease in the hydrate dissociation rate is attributed to their heat capacity and thermal conductivity. Thus, hydrate dissociation affects heat flux and thermal stability induced by the changing environmental gas composition. The effect of drilling mud on wellbore stability of hydrate sediments is largely dependent on the mineralogy and degree of sedimentation of the formation. Also, the formation porosity could affect the penetration rate of the mud filtrate which could lead to huge hydrate dissociation. Hydrate sediments with higher sand to clay ratios are more susceptible to hydrate dissociation due to their high porosity and poor consolidation with respect to their compressive strength [14].

References

1. Grigg RB, Lynes GL (1992) Oil-based drilling mud as a gas-hydrates inhibitor. SPE Drill Eng 7:32–38
2. Ebeltoft H, Yousif M, Westport I (2001) Hydrate Control during deepwater drilling : overview and new drilling-fluids formulations, 5–8
3. Lai DT, Dzialowski AK (1989) Investigation of natural gas hydrates in various drilling fluids. In: SPE/IADC drill conf new Orleans, Louisiana 18637, pp 181–194
4. Dzialowski A, Patel A, Nordbo MLLCK, Hydro N (2001) The development of kinetic inhibitors to suppress gas hydrates in extreme drilling conditions. In: Offshore mediterr. conf. exhib. Ravenna, Italy, pp 1–15
5. Mech D, Sangwai JS (2016) Effect of molecular weight of polyethylene glycol (PEG), a hydrate inhibitive water-based drilling fluid additive, on the formation and dissociation kinetics of methane hydrate. J Nat Gas Sci Eng 35:1441–1452
6. Schofield TR, Judzis A, Yousif M (1997) Stabilization of in-situ hydrates enhances drilling performance and rig safety. In: SPE Annu. tech. conf. exhib., pp 43–50
7. Yan K, Li X, Chen Z, Zhang Y, Xu C, Xia Z (2019) Methane hydrate formation and dissociation behaviors in montmorillonite. Chin J Chem Eng 27:1212–1218
8. Fereidounpour A, Vatani A (2014) An investigation of interaction of drilling fluids with gas hydrates in drilling hydrate bearing sediments. J Nat Gas Sci Eng 20:422–427
9. Zhao X, Qiu Z, Zhao C, Xu J, Zhang Y (2019) Inhibitory effect of water-based drilling fluid on methane hydrate dissociation. Chem Eng Sci 199:113–122

10. Zhao X, Qiu Z, Zhou G, Huang W (2015) Synergism of thermodynamic hydrate inhibitors on the performance of poly (vinyl pyrrolidone) in deepwater drilling fluid. J Nat Gas Sci Eng 23:47–54
11. Saikia T, Mahto V (2016) Evaluation of 1-Decyl-3-Methylimidazolium Tetrafluoroborate as clathrate hydrate crystal inhibitor in drilling fluid. J Nat Gas Sci Eng 36:906–915
12. Saikia T, Mahto V (2016) Experimental investigations of clathrate hydrate inhibition in water based drilling fluid using green inhibitor. J Pet Sci Eng 147:647–653
13. Kawamura T, Yamamoto Y, Yoon JH, Haneda H, Ohga K, Higuchi K (2002) Dissociation behavior of methane-ethane mixed gas hydrate in drilling mud fluid. Proc Int Offshore Polar Eng Conf 12:439–442
14. Zhao X, Qiu Z, Wang M, Xu J, Huang W (2019) Experimental investigation of the effect of drilling fluid on wellbore stability in shallow unconsolidated formations in deep water. J Pet Sci Eng 175:595–603

Chapter 5
Thermodynamic Behaviour of Hydrates Drilling Muds

The thermodynamic data for almost all the drilling mud tested for mitigating hydrate formation are reviewed in this section. The data provided in this section provides fundamental information on the effect of drilling mud additives on the hydrate phase equilibrium boundary data of drilling gas systems.

5.1 Phase Behavior of Hydrates in Drilling Mud

This section discusses the thermodynamic inhibition effect of gas hydrate additives in drilling mud systems. The hydrate thermodynamic drilling mud studies are mainly focused on preventing hydrate formation in deep-waters operation and not for the purposes of drilling hydrate sediments. Table 5.1 presents the reported data of the phase behavior of different drilling mud additives in varying gas systems. The main thermodynamic additives studied are electrolytes and glycols with some synergic effects for water depths above 609.6 m. The type of drilling mud additives used by authors/researchers could affect the performance of the inhibitors. The use of high salt drilling mud systems provides a very significant hydrate inhibition effect which could be further enhanced by adding other inhibitors. However, for drilling natural gas hydrates, high salt drilling mud might dissociate huge quantities of the hydrate and cause wellbore collapse and hydrate formation in the system. On the other hand, the presence of low salt concentrations and polymers have less inhibition effect on the hydrate formation with water. Maintaining such minimal inhibition effect could be useful for drilling through hydrate-bearing sediments. However, their inhibition effect should be increased for effective hydrate management in deep-water drilling operations. The presence of oil (OBMs) has a similar hydrate inhibition effect as low salts WBMs and seawater. The phase behavior effect of polymers such as XGUM, PAM, and guar gum could reduce the hydrate depression temperature up to 1 K depending on the molecular weight. However, the inhibition effect with 1 K is

© The Author(s), under exclusive license to Springer Nature Switzerland AG 2022   73
B. Lal et al., *Hydrate Control in Drilling Mud*,
SpringerBriefs in Petroleum Geoscience & Engineering,
https://doi.org/10.1007/978-3-030-94130-7_5

Table 5.1 Reported phase behavior data on drilling mud in presence of hydrate inhibitors

Author	Gas	System	T (K)	P (MPa)	ΔT (K)
Lai and Dzialowski [2]	Natural gas[3]	BM-5	294.21	8.91	Base
			294.37	10.70	Base
			292.32	6.39	Base
		BM-5 + 30 ppb gel	296.82	10.71	2.4
		BM-5 + 10 ppb salt	291.93	6.78	0.8
			293.98	9.84	0.2
			295.09	12.19	0.3
		BM-5 + 50 ppb salt	290.76	12.82	4.9
			289.82	9.89	4.4
			285.98	4.33	5.5
		BM-5 + 90 ppb salt	284.82	29.43	19.1
		BM-5 + Lignosulfonate	296.09	10.14	1.7
			290.65	3.52	0.4
			293.32	5.75	1.1
		BM-5 + Lignite	296.21	10.82	1.5
			298.43	18.89	0.3
		BM-5 + 5 ppb Gel + Caustic	294.26	7.16	1.4
		BM-5 + 20 ppb Gel	293.32	6.32	0.9
		BM-5 + 2% diesel	295.93	11.46	0.9
		Polymer A	NHF	NHF	NHF
		Polymer B	NHF	NHF	NHF
		Polymer C	287.93	33.97	18.2
		Polymer D	NHF	NHF	NHF
		Polymer E	294.54	37.85	13.5
		Sea water gyp	294.54	10.67	0.1
			294.43	11.27	0.5
			291.71	6.64	0.9
			293.43	9.29	0.5
		Lime mud	294.48	7.69	1.3
			293.65	6.55	1.1
			296.48	10.16	2.1
		BM-5 + 50 ppb glycerine	292.71	7.65	0.4
			290.32	6.67	2.3

(continued)

Table 5.1 (continued)

Author	Gas	System	T (K)	P (MPa)	ΔT (K)
		BM-5 + 50% vol glycerine	285.87	35.36	21.0
		BM-5 + 100 ppb propylene glycol	292.65	23.37	8.3
		BM-5 + 50 vol% propylene glycol	284.82	27.97	18.4
Hale and Dewan [7]	Natural gas[4]	Inhibitor Free	293.19	6.89	Base
			296.57	20.68	Base
			299.62	31.03	Base
		2,5% NaCl	290.53	6.89	2.7
			295.53	20.68	1.0
			298.50	31.03	1.1
		10% NaCl	287.09	6.89	6.1
			291.78	20.68	1.4
			294.83	31.03	1.7
		20% NaCl	280.48	6.89	19.1
			284.62	20.68	8.6
			287.67	31.03	5.5

Author	Gas	System	T (K)	P (MPa)	ΔT (K)
Hale and Dewan [7]	Natural gas[4]	10% Glycerol	289.92	6.89	6.6
			294.92	20.68	4.7
			297.89	31.03	4.7
		35% Glycerol	283.14	6.89	10.1
			287.59	20.68	9.0
			290.57	31.03	9.1
		60% Glycerol	270.08	6.89	23.1
			273.44	20.68	19.8
			276.33	31.03	20.2
		2.5% NaCl + 10% Glycerol	289.07	6.89	10.6
			293.99	20.68	0.8
			297.04	31.03	3.9
		2.5% NaCl + 35% Glycerol	281.05	6.89	15.5
			285.34	20.68	14.3
			288.32	31.03	4.9
		2.5% NaCl + 60% Glycerol	265.76	6.89	27.4
			268.73	20.68	27.8
			271.62	31.03	28.0

(continued)

Table 5.1 (continued)

Author	Gas	System	T (K)	P (MPa)	ΔT (K)
		2.5% NaCl + 10% Glycerol	285.06	6.89	8.1
			289.58	20.68	3.6
			292.63	31.03	3.9
		10% NaCl + 35% Glycerol	273.72	6.89	25.9
			277.39	20.68	15.8
			280.28	31.03	12.9
		10% NaCl + 60% Glycerol	262.86	6.89	33.7
			265.59	20.68	34.0
			268.48	31.03	24.7
		10% NaCl + 10% Glycerol	276.63	6.89	16.6
			280.53	20.68	16.0
			283.51	31.03	16.1
		20% NaCl + 35% Glycerol	269.31	6.89	23.9
			272.58	20.68	20.6
			275.56	31.03	21.0
		20% NaCl + 60% Glycerol	262.06	6.89	37.6
			264.71	20.68	28.5
			267.68	31.03	25.5
Grigg and Lynes [1]	Natural gas[5]	VC4	295.37	9.62	1.9
		BC3	294.26	8.41	0.8
		LP4	295.37	8.29	1.9
		VC3	293.71	7.58	0.2
		BC2	292.59	6.89	0.9
		VC2	294.26	6.76	0.8
			292.37	6.58	1.1
		BC1	293.15	6.38	0.3
		LP3	291.15	4.14	2.3
		VC1	290.37	3.45	3.1
		LP2	288.59	3.10	4.9
		LP1	286.93	2.54	6.5
		Oil mud + H_2O	298.43	31.85	Base
			296.76	23.10	Base
			294.54	14.82	Base
		VC3	291.98	9.03	1.5
		VC2	290.26	6.93	3.2
			288.15	6.55	5.3

(continued)

Table 5.1   (continued)

Author	Gas	System	T (K)	P (MPa)	ΔT (K)
			286.48	4.14	7.0
		VC1	283.98	3.46	9.5
		28.83 wt% CaCl₂ brine	NHF	NHF	NHF

Author	Gas	System	T (K)	P (MPa)	ΔT (K)
Grigg and Lynes [1]	Natural gas[5]	19.22 wt% CaCl₂ brine	287.87	27.72	5.6
			287.21	23.10	6.3
			285.54	19.58	7.9
			284.82	19.44	8.6
			283.43	12.89	10.0
			282.04	6.89	11.4
			281.04	6.89	12.4
		14.42 wt% CaCl₂ brine	289.82	14.34	3.6
			287.82	7.10	5.6
		2.8 wt% CaCl₂ brine	293.15	7.58	0.3
		Oil mud + 19.22 wt% CaCl₂ brine	284.26	28.96	9.2
			282.04	19.65	11.4
			278.15	12.41	15.3
			275.93	5.79	17.5
			274.26	4.00	19.2
Kotkoskle et al. [3]	Natural gas[6]	NOPH	286.21	25.79	0.6
			288.09	20.89	3.2
			283.09	14.00	0.8
			281.98	12.48	1.7
			283.98	19.93	0.7
		SA12	293.76	27.85	7.9
			291.76	22.27	6.7
			291.21	14.41	7.3
			292.54	21.37	7.6
		Mud1	290.98	2.72	8.7
			293.15	4.14	10.7
			293.76	5.52	11.1
			294.87	7.10	12.0
			300.93	26.75	15.2
			300.93	27.17	15.2
			294.26	5.71	11.6

(continued)

Table 5.1 (continued)

Author	Gas	System	T (K)	P (MPa)	ΔT (K)
			301.26	28.82	15.3
			301.76	28.37	15.8
		BRMU	283.71	27.23	2.1
			283.43	20.62	1.4
			283.71	15.38	0.4
			280.82	11.03	2.6
			281.54	16.51	2.7
			284.87	26.23	0.7
			283.43	20.34	1.4
		SAT	279.82	23.44	5.4
			277.32	17.82	7.1
			280.09	25.79	5.5
			279.15	20.34	5.6
			280.54	27.30	5.2
			277.48	19.86	7.2
			277.87	19.58	6.8
		BR20	291.93	27.85	6.1
			290.98	21.65	6.0
			288.43	13.65	4.6
			290.71	20.51	5.9
			291.43	27.03	5.7
			288.26	14.41	4.3
		NOBB	285.87	25.79	0.3
			283.71	17.10	0.6
			281.82	32.20	4.7
Author	Gas	System	T (K)	P (MPa)	ΔT (K)
Kotkoskle et al. [3]	Natural gas[6]	NOBE	287.09	26.13	1.5
			286.21	21.79	1.2
			284.87	16.13	0.7
			285.87	21.37	0.9
			283.26	12.34	0.4
		NACL	293.93	27.96	8.1
			281.54	17.93	2.9
			285.32	25.34	0.2
			281.65	12.48	2.0
			283.43	18.55	1.1

(continued)

Table 5.1   (continued)

Author	Gas	System	T (K)	P (MPa)	ΔT (K)
			284.87	26.03	0.7
		BBM	285.87	27.51	Base
			285.32	20.96	Base
			284.59	16.58	Base
			285.43	27.94	Base
			283.09	11.48	Base
		NOXC	285.15	28.51	0.8
			283.71	20.86	1.2
			281.54	10.48	1.8
			284.87	20.68	0.0
		Mud2	290.98	28.96	5.0
			287.98	21.17	3.1
			286.48	12.31	2.8
			287.54	19.34	2.9
			290.09	27.92	4.2
		BR30	283.43	26.23	2.2
			281.98	20.62	2.8
			280.98	12.20	2.6
			283.09	23.65	2.2
			284.26	19.86	0.5
			282.54	13.24	1.2
Yousif and Young [8]	Natural gas[7]	5 wt% $CaCl_2$ + 5 wt% Glycerol	295.65	17.95	0.5
			296.76	23.10	0.4
		10 wt% $CaCl_2$ + 10 wt% Glycerol	292.15	18.17	4.0
			291.48	12.96	3.7
		5 wt% KCl + 5 wt% Glycerol	296.26	18.96	0.1
			294.15	11.34	0.7
		10 wt% KCl + 10 wt% Glycerol	293.15	19.34	3.3
			291.71	13.20	3.5
		3 wt% NaCl + 5 wt% KCl + 5 wt% Glycerol	294.15	19.93	2.4
			292.71	15.72	3.0
		10 wt% NaCl + 10 wt% KCl + 10 wt% Glycerol	285.37	12.49	9.7
			284.26	21.65	12.6
			283.71	17.51	12.3
		5 wt% NaCl + 5 wt% KCl + 15 wt% Glycerol	289.98	13.38	5.3
			289.04	18.20	7.1

(continued)

Table 5.1 (continued)

Author	Gas	System	T (K)	P (MPa)	ΔT (K)
			288.43	14.62	7.1
		20 wt% NaCl + 20 wt% Glycerol	274.71	27.89	23.4
			274.04	20.99	22.7
		30 wt% Glycerol	292.04	14.10	3.3
			290.37	14.38	5.1
			288.98	28.99	9.3

Author	Gas	System	T (K)	P (MPa)	ΔT (K)
Yousif and Young [8]	Natural gas[7]	20 wt% Glycerol	294.43	21.39	2.4
			296.93	28.54	1.3
			296.21	22.34	0.8
			297.59	28.20	0.5
			295.93	21.13	0.8
			295.21	22.48	1.8
			293.43	14.56	2.0
		10 wt% Glycerol	296.59	27.96	1.5
			294.15	15.27	1.5
			297.32	13.79	2.0
			294.26	27.96	3.8
			293.87	17.93	2.3
		20 wt% NaCl	293.93	25.34	3.6
			281.54	12.48	13.5
			285.32	25.17	12.2
			281.65	20.89	15.1
		30.6 wt% NaBr	283.43	13.93	11.9
			282.87	27.58	15.1
			282.48	21.37	14.3
		20 wt% NaCl + 10 wt% Glycerol	282.87	11.86	12.1
			279.87	28.44	18.3
			278.71	18.53	17.5
		20 wt% NaCl + 20 wt% Glycerol	274.54	14.41	20.9
			273.21	27.85	24.8
			272.59	21.65	24.3
		20 wt% NaBr	291.93	13.65	3.4
			290.98	36.49	8.7
			288.43	20.11	8.1
		20,1wt% NaCl + 20wt% Glycerol	277.15	35.36	22.4
			273.54	18.55	22.7

(continued)

Table 5.1 (continued)

Author	Gas	System	T (K)	P (MPa)	ΔT (K)
		22.1 wt% NaCl + 15 wt% Glycerol	277.26	37.38	22.6
			275.54	20.43	21.1
		23.4 wt% NaCl + 10 wt% Glycerol	279.87	35.44	19.7
			277.26	35.58	22.3
		26 wt% NaCl	282.59	25.39	15.0
			283.04	20.05	13.5
			280.43	20.29	16.2
			279.98	41.37	20.7
		15,1 wt% CaCl$_2$	290.37	41.27	10.3
			294.82	39.84	5.6
			290.98	20.76	5.7
		30 wt% Glycerol	297.59	41.93	3.2
			290.76	12.82	4.4
		20 wt% Glycerol	289.82	9.89	4.8
			285.98	4.33	7.5
		20,44 wt% CaCl$_2$	284.82	29.43	13.5
		14,42 wt% CaCl$_2$	289.82	14.34	5.6
			287.82	7.10	6.2
		19,22 wt% CaCl$_2$	287.87	27.72	10.2
			287.21	23.10	9.9
			285.54	19.58	10.9
			284.82	19.44	11.6
			283.43	12.89	11.7
			282.04	6.89	11.9
			281.04	6.89	12.9
Author	Gas	System	T (K)	P (MPa)	ΔT (K)
Ebeltoft et al. [4]	Natural gas[8]	Sea water	299.82	37.92	Base
			297.82	26.34	Base
			297.32	20.79	Base
			296.43	20.62	Base
			294.37	10.33	Base
		BM (20 wt% NaCl + 10%AQUA-COL S)	280.93	36.96	Base
			278.82	25.03	Base
			277.87	19.58	Base
			275.21	10.41	Base

(continued)

Table 5.1 (continued)

Author	Gas	System	T (K)	P (MPa)	ΔT (K)
		20 wt% NaCl + 20 wt% GEO-MEG D207	279.37	27.13	0.3
			277.48	20.45	0.2
			275.32	10.55	0.3
			281.48	36.78	0.4
		20 wt% NaCl + 10 wt%HF-100 N	280.32	34.94	0.4
			278.93	24.55	0.4
			277.82	18.89	0.5
			275.71	10.48	0.1
		20 wt% NaCl + 10 wt%EG	278.54	33.37	1.9
			277.87	27.30	1.2
			276.26	19.69	1.2
			274.26	11.33	1.5
		20 wt% KCl	292.09	38.47	10.6
			289.09	27.14	10.0
			287.93	20.28	10.3
			287.04	11.60	11.2
		10 wt% KCl + 10%AQUA-COL S	294.93	36.16	13.9
			293.21	26.72	14.2
			291.98	20.48	14.3
			272.04	10.29	3.5
		10 wt% KCl + 10 wt%HF-100 N	294.37	37.65	13.1
			292.21	25.92	13.4
			290.26	19.72	12.7
			288.48	10.60	12.9
		10 wt% KCl + 10 wt% NaCl	289.59	34.03	9.1
			288.04	26.71	9.0
			286.04	19.58	8.6
		10 wt% KCl + 10 wt% NaCl	283.26	10.27	7.7
		10 wt% KCl + 10 wt% NaCl + 10%AC	283.71	33.27	3.3
			282.32	23.76	3.9
			281.09	19.65	3.6
			279.26	9.86	3.8
		10%AQUA-COLS	299.37	37.92	18.0
			299.59	27.58	20.4
			296.37	19.83	18.8

(continued)

Table 5.1   (continued)

Author	Gas	System	T (K)	P (MPa)	ΔT (K)
			294.76	10.49	19.2
		20 wt% Na-formate + 10%AQUA-COLS	283.48	31.51	3.5
			282.43	25.92	3.6
			281.21	19.58	3.7
			279.21	10.32	3.7
		40 wt% Na-formate	278.71	30.34	1.1
			278.48	32.34	1.7
			277.09	24.42	1.4
			277.04	19.86	0.5
			275.21	11.31	0.5
		80:20 SYN-TEQ with 30% CaCl$_2$	NHF	NHF	NHF
			NHF	NHF	NHF
Author	Gas	System	T (K)	P (MPa)	ΔT (K)
Ebeltoft et al. [4]	Natural gas[8]	9.33 wt% lignite with 1.86% NaOH	298.15	35.30	17.3
			296.98	27.23	17.9
			295.59	20.44	17.9
			293.15	10.05	17.7
		15 wt% KCl + 30 wt%HF-100 N	283.09	30.89	3.2
			282.65	25.65	3.9
			281.54	19.51	4.1
			279.59	10.24	4.1
		21 wt% Amonium calcium nitrate	297.48	36.65	16.4
			295.37	26.30	16.5
			294.26	20.17	16.7
			291.71	10.48	16.1
		80:20 SYN-TEQ mud w/30% CaCl$_2$	NHF	NHF	NHF
			NHF	NHF	NHF
		5 wt% KCl + 15 wt% NaCl + 10%AC mud	283.87	32.41	3.7
			283.15	25.03	4.5
			281.21	18.62	3.9
			279.26	9.58	3.9
		5 wt% KCl + 15 wt% NaCl mud	286.37	33.09	6.0
			285.43	24.55	6.9
			284.26	19.72	6.7
			282.59	11.95	6.7
		5 wt% KCl + 15 wt% NaCl + 10%EG mud	282.59	33.78	2.1
			280.93	27.75	1.7

(continued)

Table 5.1 (continued)

Author	Gas	System	T (K)	P (MPa)	ΔT (K)
			280.26	20.44	2.6
			279.26	10.38	3.7
		23.8 wt% Na-formate + 28.6 wt%EG	NHF	NHF	NHF
			NHF	NHF	NHF
		40 wt% Na-formate mud	287.59	31.54	7.6
			286.76	26.20	7.9
		40 wt% Na-formate mud	285.21	18.96	7.9
			283.21	9.27	7.9
		23 wt% NaBr + 25 wt% EG	274.26	28.44	5.1
			273.15	23.10	5.1
			272.59	18.62	4.7
			271.15	9.98	4.3
		80:20 SYN-TEQ w/15 wt% CaCl$_2$	289.98	36.37	8.9
			288.71	26.20	9.8
			286.76	19.99	9.2
			284.98	10.41	9.4
Dalmazzone et al., [6]	CH$_4$	CaCl2 10 wt%	277.95	6.00	3.6
			279.85	8.01	4.0
			281.35	10.00	4.7
			278.35	6.88	4.2
			279.95	7.78	3.6
			282.25	10.29	4.2
		CaCl2 20 wt%	270.05	6.33	11.9
			271.25	7.45	11.9
			272.25	8.56	12.2
			273.55	10.29	12.9
		CaCl2 23 wt%	263.95	6.20	17.8
			266.45	7.90	17.3
			267.35	9.00	17.6
			267.75	9.95	18.3
Gupta et al. [9]	CH$_4$	PAM1-50 ppm	283.00	8.17	1.0
			282.40	7.47	0.8
			281.60	6.76	0.8
			280.50	6.16	1.2

Author	Gas	System	T (K)	P (MPa)	ΔT (K)
Gupta et al. [9]	CH$_4$	PAM1-100 ppm	283.50	8.31	0.7

(continued)

Table 5.1 (continued)

Author	Gas	System	T (K)	P (MPa)	ΔT (K)
			282.60	7.57	0.7
			280.10	5.75	1.2
			280.90	6.40	1.1
		PAM1-200 ppm	282.80	7.77	0.8
			282.00	7.05	0.7
			280.80	6.18	1.0
		PAM1-500 ppm	284.60	8.56	0.2
			284.00	8.10	0.1
			282.20	6.80	0.3
			281.40	6.20	0.4
		PAM2-100 ppm	283.40	7.94	0.3
			282.50	7.39	0.6
			281.80	6.76	0.6
			280.40	5.90	1.0
		PAM2-200 ppm	282.90	7.85	0.7
			281.90	7.15	1.0
			281.30	6.62	1.0
			280.80	6.30	1.1
		PAM2-500 ppm	283.40	8.06	0.5
			282.70	7.54	0.6
			281.20	6.54	1.0
		XG1-100 ppm	283.80	8.63	0.7
			283.00	7.83	0.6
			281.80	6.78	0.6
		XG1-200 ppm	283.40	7.89	0.3
			283.00	7.54	0.3
			281.50	6.39	0.5
			279.80	5.50	1.2
		XG1-500 ppm	283.40	7.81	0.2
			282.40	7.10	0.4
			281.30	6.34	0.6
		XG2-100 ppm	283.80	8.62	0.7
			282.90	7.81	0.7
			281.60	6.76	0.8
		XG2-200 ppm	283.30	7.93	0.4
			282.80	7.45	0.4
			282.10	6.83	0.4
			280.90	6.10	0.8

(continued)

Table 5.1 (continued)

Author	Gas	System	T (K)	P (MPa)	ΔT (K)
		XG2-500 ppm	283.00	7.81	0.6
			282.20	7.17	0.7
			281.30	6.57	0.9
			280.80	6.18	1.0
		GG1-100 ppm	283.40	7.93	0.3
			282.80	7.48	0.4
			281.40	6.70	0.9
		GG1-200 ppm	283.20	8.05	0.7
			281.50	6.90	1.1
			280.90	6.46	1.2
			280.30	6.01	1.3
		GG1-500 ppm	283.30	7.56	0.0
			282.50	7.00	0.2
			281.80	6.49	0.3
			280.40	5.67	0.8
		GG2-100 ppm	283.30	8.07	0.6
			282.20	7.10	0.6
			281.60	6.60	0.6
			280.00	5.83	1.4
Gupta et al. [9]	CH_4	GG2-200 ppm	283.40	8.13	0.6
			281.90	7.08	0.9
			281.40	6.60	0.8
			280.70	6.15	1.0
		GG2-500 ppm	282.90	7.99	0.9
			281.80	7.00	0.9
			281.00	6.34	0.9
			280.20	5.88	1.2

[4]$(87.29C_1 + 7.55C_2 + 3.09C_3 + 0.49i\text{-}C_4 + 0.79n\text{-}C_4 + 0.39i\text{-}C_5 + 0.4N_2)$

[5]$(87.55C_1 + 7.59C_2 + 3.1C_3 + 0.19i\text{-}C_4 + 0.79n\text{-}C_4 + 0.19i\text{-}C_5 + 0.19n\text{-}C_6 + 0.39N_2)$; [6]$(87.2C1 + 7.6C2 + 3.1C3 + 0.5i\text{-}C4 + 0.8n\text{-}C4 + 0.2i\text{-}C5 + 0.2n\text{-}C5 + 0.4N_2)$; NOPH (349 g H20, 82 g NaCl, 13,75 g bentonite, 0,63 g XCD, 205,4 g Barite); SA12 (349 g H20, 42 g NaCl, 13,75 g bentonite, 0,63 g PHPA, 0,63 g XCD, 205,4 g Barite)

Mud1 (335,6 g H20,12 g bentonite, 1 g PHPA,20 g drill solids, 0,8 g NaOH, 2 g Q-Broxin); BRMU (349 g H20, 147,9 g NaCl-20%, 13,75 g bentonite, 0,63 g PHPA, 0,63 g XCD, 205,4 g Barite); SAT (349 g H20, 118,5 g NaCl, 13,75 g bentonite, 0,63 g PHPA, 0,63 g XCD, 205,4 g Barite); NOBB (349 g H20, 82 g NaCl, 0,63 g PHPA, 0,63 g XCD); NOBE (349 g H20, 82 g NaCl, 0,63 g PHPA, 0,63 g XCD, 205,4 g Barite); [7]$(87.26C1 + 7.57C2 + 3.1C3 + 0.49i\text{-}C4 + 0.79n\text{-}C4 + 0.2i\text{-}C5 + 0.2n\text{-}C5 + 0.39N_2)$; BBM (349 g H20, 82 g NaCl-20%, 13,75 g bentonite, 0,63 g PHPA, 0,63 g XCD, 205.4 g Barite); NOXC (349 g H20, 82 g NaCl, 13,75 g bentonite, 0,63 g PHPA, 205.4 g Barite); Mud2 (287,3 g H20, 50,3 g NaCl, 1 g PHPA, 1 g XCD, 20 g drill solids, 150 g Barite)

[8]$(87.243C1 + 7.57C2 + 3.08C3 + 0.51i\text{-}C4 + 0.792n\text{-}C4 + 0.202i\text{-}C5 + 0.2n\text{-}C5 + 0.403N_2)$; AQUA-COLS (AC)

relatively negligible when dealing with deep-water hydrate management. But could be somewhat significant in hydrate-bearing sediment drilling operations.

Grigg and Lynes [1] reported that oil and continuous phase with 20 vol% water in OBM shifts the hydrate phase curve to the left by decreasing the gas-hydrates formation temperature from 258.15 to 260.93 K within 3.45 to 31.72 MPa. However, the presence of oil (diesel ~ 2 wt%) as a non-continuous phase has very little to negligible effect on the hydrate phase boundary curve [2]. The presence of dissolved solids in the oil-based mud has the tendency of decreasing the hydrates formation temperature. On the other hand, some WBM additives (PHPA, XGUM, and bentonite) are known to slightly promote hydrate formation [3]. $CaCl_2$ concentration above 25 wt% completely prevents hydrate formation, especially in OBMs [1, 4]. Interestingly, other drilling mud additives have a negligible effect on the THI effect of $CaCl_2$ in drilling mud systems [5, 6]. However, the inhibition effect of $CaCl_2$ decreases with decreasing concentration [5, 6]. For example, at 6.9 MPa 19.22 wt% $CaCl_2$ reduced the natural gas hydrate phase behavior from 267.03 K to 1 °F at 2.88-wt% $CaCl_2$ [1]. Hence, confirming that hydrate formation in emulsion fluids is controlled mainly by the salt content in the water phase. The synergy of sodium formate brine and ethylene glycol at high concentrations (23.8 wt% Na-formate and 128.6 wt% glycols) completely prevents hydrate formation. Hence, the use of salts at very high concentrations has a very strong hydrate inhibition effect [4].

The effect and impact of other classes of salts on hydrate formation in drilling mud show that most deep-water drilling operations (up to 2286 m) employ NaCl/polymer systems (20–23 wt% NaCl). This is because NaCl has proven to be the most efficient salt inhibitor on the basis of mass fraction [1]. Another reason NaCl is mostly used is because of its compatibility with standard mud products over other salts [2]. However, salts could promote rapid corrosion when used. The molecular weight of salts, valency, and degree of ionization controls their thermodynamic hydrate inhibitory effect. Salts are effective hydrate inhibitors and their inhibitive effect is KCl next to NaCl, followed by $CaCl_2$, NaBr, Na-formate, and calcium nitrate [1]. Na-formate has good inhibition properties especially due to its solubility in water. But they crystalize at high concentration (40 wt.%) which limits their application [1]. Sea-water gyp based-muds can reduce the hydrate equilibrium temperature like other salts present its composition.

However, lime mud has no significant effect on hydrate phase boundary [2]. On the other hand, ethylene glycol best inhibits gas hydrate formation than polyglycol, making EG a superior hydrate inhibitor amongst the glycols. Ethylene glycol is a better inhibitor because more hydroxyl groups are available to make hydrogen bonds with water molecules, and hence, it is more difficult for water molecules to participate in the hydrate structure.

Glycerin and propylene glycol also has a strong inhibition impact on the hydrate equilibrium temperature, especially at relatively high concentrations for desire inhibition in deep-water operations [2]. A synergic study between glycols and salts by Ebeltoft et al. [1] shows a good gas hydrate inhibition with a relatively low combination (5–15 wt%). They proposed that mud from 5 wt% KCl + 15 wt% NaCl 10 wt% ethylene glycol was optimum to allow about 290.15 K hydrate suppression temperature.

Hale and Dewan [7] confirmed the excellent hydrate inhibition synergy between salt and glycol using NaCl and glycerol. Significant synergy is observed at high concentrations, with poor or no synergy at low concentrations. This means that the synergy between salts and glycerol is concentrations dependent [7]. Salts concentrations below 5 wt% and 10 wt% glycerol will have a poor or negligible hydrate inhibition effect. However, to effectively optimize their synergy the salt and glycerol concentration should not exceed 25 wt% and 40 wt%, respectively [7]. Lai and Dzialowski [2] indicated that lignosulfonate and lignite have negligible inhibition effects on gas hydrate phase behavior. They show a slight thermodynamic promotional effect at standard mud-designed concentrations. Similarly, Caustic and gel mud systems do not affect hydrate equilibrium boundary curves, when used at normal field mud design concentrations.

Some conventional drilling mud polymers (PAM, GG, and XGUM) have been reported to exhibit slight hydrate inhibition efficiency. They can suppress the hydrate equilibrium temperature up to 1 K with 0.01–0.05 wt%. They showed better inhibition at low concentrations and low molecular weight [9]. It is argued that the activity of polymers to perturb water structure is enhanced via higher mobility solution, owing to their low molecular weight. Also, the presence of hydroxyl groups ($-OH$) in the polymer structure further enhances their perturbation activity on water molecules [9]. In contrary, Kotkoskie et al. [3] argued that, XGUM and PHPA promote the phase behavior of hydrate formation. This means that they slightly increase the hydrate formation temperature. Perhaps the disagreement might arise from the molecular weight of the XGUM used or maybe the source of the XGUM extracts. The phase behavior studies of drilling muds are mainly focused on deepwater drilling operations, more studies on the hydrate equilibrium of mud formulated for drilling hydrate sediments should be tested and discussed in literature. In addition, drilling fluids hydrate-based testing should be conducted in mass fraction and not in other concentration units [10]. This is because mass fractions are mostly used in the industry and could provide a better comparison of additives.

References

1. Grigg RB, Lynes GL (1992) Oil-based drilling mud as a gas-hydrates inhibitor. SPE Drill Eng 7:32–38
2. Lai DT, Dzialowski AK (1989) Investigation of natural gas hydrates in various drilling fluids. In: SPE/IADC drill conf New Orleans, Louisiana 18637, pp 181–194
3. Hale H, Dewan AKR (1990) Inhibition of gas hydrates in deepwater drilling. SPE Drill Eng 7:109–115
4. Ebeltoft H, Yousif M, Westport I (2001) Hydrate control during deepwater drilling: overview and new drilling-fluids formulations 5–8
5. Herzhaft B, Dalmazzone C (2001) Gas hydrate formation in drilling MUD characterized with DSC technique. In: SPE Annu. tech. conf. exhib. New Orleans, Louisiana, pp 575–584

6. Dalmazzone D, Dalmazzone C, Herzhaft B (2002) Differential scanning calorimetry: a new technique to characterize hydrate formation in drilling muds. SPE J 7:196–202

7. Hale AH, Dewan AKR (1990) Inhibition of gas hydrates in deepwater drilling. SPE Drill Eng 5(109–115):18638

8. Yousif MH, Young DB (1993) Simple correlation to predict the hydrate point suppression in drilling fluids. Drill Conf—Proc 287–294

9. Gum X, Gum G, Gupta P, Nair VC, Sangwai JS (2019) Phase equilibrium of methane hydrate in aqueous solutions of polyacrylamide, xanthan gum, and guar gum. J Chem Eng Data 64:1650–1661

10. Bavoh CB, Lal B, Osei H, Sabil KM, Mukhtar H (2019) A review on the role of amino acids in gas hydrate inhibition, CO_2 capture and sequestration, and natural gas storage. J Nat Gas Sci Eng 64:52–71

Chapter 6
Hydrates Drilling Muds Rheological Properties

This section discusses the drilling mud rheological properties data suitable for drilling hydrate sediment. Almost all the articles related to hydrate drilling mud for drilling hydrate sediment and hydrate management in drilling during deepsea drilling operations are discussed. The gathered data is tabulated for academic and practical field use. Drilling mud properties such as PV, AV, FL, and gel strength are the main focus of this section.

6.1 Rheology Data of Hydrates in Drilling Mud

Table 6.1 presents the reported data of the rheology of different drilling mud additives in varying drilling mud systems. Saikia et al. [1] Reported that at lower concentration ethylene glycol showed no considerable effect on the rheology of drilling fluid. The considerable variation in the rheological characteristic of drilling mud was mainly because of the temperature drop. Overall, carbon dots (CDs) showed a marginal effect on the rheological characteristics of drilling fluid in comparison to ethylene glycol. The effect of the concentration of CDs (below 1.0 wt%) on the drilling fluid rheology was negligible. ethylene glycol and CDs show a very marginal effect on the plastic viscosity of drilling fluid. Nikolaev et al. [2] also reported that polyethylene glycol-based mud's density, YP, PV, and Gel value have increasing trends as the temperature decreases, but the changes are small and acceptable. Furthermore, the high ratios of yield point and plastic viscosity will help to carry cuttings and clean the borehole. Although filtration has an increasing trend, it shows very little change. Polyethylene glycol exhibits good performance on the basis of its rheological properties and stability at low temperatures. The effect of PVP K90 on the rheological properties of polyethylene-glycol drilling fluid provides a gel strength and yield point of drilling fluid that is preferable when the concentration of PVP K90 increases to 1%. The yield point/plastic viscosity ratio approaches 0.48, which

© The Author(s), under exclusive license to Springer Nature Switzerland AG 2022   91
B. Lal et al., *Hydrate Control in Drilling Mud*,
SpringerBriefs in Petroleum Geoscience & Engineering,
https://doi.org/10.1007/978-3-030-94130-7_6

also shows that the drilling fluid has good rheological properties and is capable of suspending drilling cuttings and stabilizing boreholes. If the concentration of PVP K90 increases more than 1%, the rheological properties of drill fluid become worse, which will influence the circulation rate of drilling fluid and even cause pressure surges in the borehole. Therefore, the concentration of 1% PVP K-90 can achieve the balance of good hydrate inhibition and relatively suitable rheology behavior in the polyethylene-glycol drilling fluid (Table 6.1).

Zhao et al. [3] claimed that the addition of PVP has a significant influence on the rheological property of the drilling fluid at 2 °C. The gel strength (Gel), yield point, and viscosity of the drilling fluid substantially increased following the addition of 0.1 wt% PVP. Increasing the concentration of PVP could lead to an increase in rheological parameters; the rheological parameters of the drilling mud could be increased by two and three times when 0.5 wt% and 1.0 wt % PVP were added, respectively. Filtration loss was reduced as the concentration of PVP was increased. Since polymers could increase the viscosity of the filtrate and thus help form the filter cake along the borehole walls. In deepwater drilling operations, high rheology of the drilling fluid is not favorable because it can lead to high equivalent circulating density, causing severe lost circulation. Zhao et al. [16] also suggested that PVP increases the PV and YP of drilling mud with and without NaCl (Salts). The rheological properties of Polyacrylate fluid and advanced polymer mud do not have a significant difference in PV, YP, and filtration properties [6]. In comparison with PVP, lecithin is a type of surfactant and has a much smaller impact on the rheological properties of the drilling fluid. As the concentration of lecithin increase from 0.1 to 1.0 wt%, the rheological parameters of the drilling fluid slightly increase, and the filtrate loss was slightly reduced. The addition of 1.0 wt% lecithin increases the PV of the drilling fluid from 15 to 22 mPa s and increases the YP from 8 to 9 Pa. Therefore, the addition of 0.1–1.0 wt% lecithin would not significantly impact the rheological and filtration properties of the drilling fluid, although its foaming characteristics present a serious problem at higher concentrations.

According to Wang et al. [11] the thickening effect of CMC, XGUM, and MS, in drilling mud, is in the order of CMC > XGUM > MS. This is due to the different aggregation degrees of their formed network skeletons aggregated in the thickener-hydrated molecule solution or drilling mud. The rheological properties of MS and XGUM solutions are more thermally stable at low temperatures than CMC [11]. Temperature is claimed to have more effect on the rheology of drilling fluid than the concentration of hydrate inhibitors based on statistical optimization analysis [17]. Similar findings were confirmed by Jiang et al. [18]. The mixture of ethylene glycols and PVP could yield efficient drilling mud rheology suitable for drilling. The plastic viscosity and yield stress of drilling fluid slightly increase with decreasing temperature in clay-free polyglycol drilling. Thus, the response to temperature changes of clay affects drilling mud adversely. Also. clay is an important factor for the increase of plastic viscosity and yield point of drilling fluid under low temperature. The main reason is that under low temperature, the diffusion ability of cations in the diffuse layer of clay particle surface reduces, and zeta potential falls. The infiltration ability of water molecules into clay inner reduces, and the dispersion degree of clay particles decreases, thereby the friction among clay particles increases, which causes the plastic viscosity to increase. Clay particles are prone to form a comparatively strong space grid in end-end and face-face structure, which causes the yield value to increase.

Srungavarapu et al. [4] reported that CMC-based drilling fluid obeys the power-law, however, alongside xanthan gum it showed Herschel-Bulkley nature. With the decrease in temperature, the mud with CMC and XGUM causes an increase in yield stress, plastic viscosity, yield point, apparent viscosity, and flow index. Similarly, with an increase in pressure, the drilling mud showed an increase in yield stress, plastic viscosity, yield point, apparent viscosity, and a decrease in flow index. Generally, drilling mud gel strength, apparent viscosity, yield point, and plastic viscosity increases with increasing calcium and magnesium ions concentrations above their standard seawater concentration [4]. Similarly, Jiang et al. [10] showed that the shear viscosity of drilling fluid increases with increasing amounts of $MgCl_2$ and $CaCl_2$, but the changes are very small, as was the filtration. Grigg and Lynes [14] also supported such findings by reported that 19.22-wt% $CaCl_2$ increases the PV and YP of OBM but has a negligible effect on its gel strength. Interestingly, the viscosity of XGUM drilling fluid is greater than the drilling mud system containing KCl [8]. Meanwhile, Zhang et al. [9] claim that NaCl increases drilling mud YP but decreases its PV, while EG shows the opposite. However, they both have a negligible effect on mud filtrate. Glycerol has no significant impact on mud properties other than improving its fluid loss [13].

Tinku and Mahto [12] claimed that the gel strength of the drilling fluid system remains approximately the same for all the concentrations of LDHIs. Overall, pancreatic lipase shows a very small variation in rheological characteristics of base drilling fluid in comparison to PVP and hence proved to be effective and convenient to be used in drilling fluid. Liyi et al. [7] showed that, with increasing concentration of four additives, all density ρ, Marsh funnel viscosity ηm, plastic viscosity PV, and apparent viscosity AV have increasing trends. The rheological parameters of additive HT are higher at the same concentration than the parameters of the other three additives (SMK, FCLS, and SHR). These results indicate that the additive HT has better performance of stability and rheological behavior at low temperatures.

ILs have been recently reported as potential additives that can improve the rheological and filtration properties of WBMs for HPHT wells at low concentrations due to their high thermal stability and cation exchange ability [19, 20]. A recent article also shows that ILs can be used to suppress hydrate formation in drilling hydrate sediments. Tinku and Vikas [21] found that 1-Decyl-3-Methylimidazolium Tetrafluoroborate could effectively inhibit hydrate formation better than PVP in drilling fluids. Organic salts have been proposed to have the potential of reducing the strength of the downhole rock surface at low temperatures, which is conducive to improving the drilling rate in hydrate-bearing sediments. Currently, there is very limited work on the influence of ILs on drilling mud in open literature. In addition, the currently available literature has shown an undesired impact on drilling muds. Therefore, due to the huge ILs database coupled with the potentials of ILs to act as good additives

Table 6.1 Reported rheology data on drilling mud in presence of hydrate inhibitors

Author	System	AV (mPa.S)	PV (mPa.S)	YP (Pa)	Gel (Pa)	Fl (ml)	YP/PV (Pa/mPa.S)
Zhao et al. [3]	BM-8	23	15	8	1.5/2.5	6.2	0.53
	0.1 wt% lecithin	23.5	15	8.5	2/3.5	5.6	0.57
	0.25 wt% lecithin	26	18	8	2/3.5	5.7	0.44
	0.5 wt% lecithin	27.5	19	8.5	2/4	5.7	0.45
	0.75 wt% lecithin	28.5	20	8.5	2/3.5	5.4	0.43
	1 wt% lecithin	31	22	9	2/3.5	5.2	0.41
	0.1 wt% PVP	32	22	10	3/5	5.8	0.45
	0.25 wt% PVP	41.5	28	13.5	4/6.5	5.1	0.48
	0.5 wt% PVP	58	36	22	5/7.5	4.9	0.61
	0.75 wt% PVP	61	37	24	5/7.5	4.6	0.65
	1 wt% PVP	72	46	26	5.5/8.5	4.1	0.57
Srungavarapu et al. [4]	BM-9	10	8	2.4	0.48/0.67	9.8	0.30
	0.4 wt% CMC	26.5	17	9.1	0.5/1.68	7.8	0.54
	0.3 wt% XG	26	12	13.4	4.3/5.74	9.3	1.12
	0.1[a] CMC + 0.4 [a] XG	40	18	21.06	8.14/91	8.5	1.17
	0.3 [a] CMC + 0.4 [a] XG	54.5	24	29.2	9.1/10.5	8.1	1.2
	0.1 [a] CMC + 0.6 [a] XG	64	23	39.2	14.36/17.2	8.4	1.7
Fereidounpour and Vatani [5]	BM-10	NA	10	11.97	4.79/7.18	15	1.20
	BM-11	NA	10	7.18	2.39/ 3.35	10	0.72
	BM-12	NA	14	8.14	2.87/3.83	7	0.58
	BM-13	NA	15	8.62	2.87/3.83	6	0.57

(continued)

Table 6.1 (continued)

Author	System	AV (mPa.S)	PV (mPa.S)	YP (Pa)	Gel (Pa)	Fl (ml)	YP/PV (Pa/mPa.S)
Fereidounpour and Vatani [6]	BM-14	NA	14	9.58	3.83/5.74	5	0.68
	BM-15	NA	15	8.62	2.87/3.83	6	0.57
Liyi et al. [7]	BM-16	NA	NA	MFV	NA	NA	NA
	0.2 wt% HT	27.97	21.96	25.99	NA	24	NA
	0.5 wt% HT	30.03	22.95	29.94	NA	20	NA
	0.8 wt% HT	31.90	24.93	35.00	NA	18	NA

Author	System	AV (mPa.S)	PV (mPa.S)	YP (Pa)	Gel (Pa)	Fl (ml)	YP/PV (Pa/mPa.S)
Liyi et al. [7]	1 wt% HT	33.84	25.97	37.92	NA	16	NA
	1.2 wt% HT	34.84	27.95	39.98	NA	10	NA
	1.5 wt% HT	35.82	29.94	47.96	NA	8	NA
	2 wt% HT	38.75	31.96	50.97	NA	9	NA
	0.2 wt% FCLS	22.01	19.98	27.89	NA	NA	NA
	0.5 wt% FCLS	27.80	21.01	29.94	NA	NA	NA
	0.8 wt% FCLS	27.98	21.01	31.84	NA	NA	NA
	1 wt% FCLS	27.76	21.91	33.89	NA	NA	NA
	1.2 wt% FCLS	29.77	22.91	34.84	NA	NA	NA
	1.5 wt% FCLS	32.78	23.94	35.87	NA	NA	NA
	2 wt% FCLS	34.76	24.98	38.95	NA	NA	NA
	0.2 wt% SMT	20.05	12.95	24.41	NA	NA	NA
	0.5 wt% SMT	25.03	13.98	24.80	NA	NA	NA

(continued)

Table 6.1 (continued)

Author	System	AV (mPa.S)	PV (mPa.S)	YP (Pa)	Gel (Pa)	Fl (ml)	YP/PV (Pa/mPa.S)
	0.8 wt% SMT	25.95	14.93	25.99	NA	NA	NA
	1 wt% SMT	28.84	15.97	26.62	NA	NA	NA
	1.2 wt% SMT	31.73	18.08	30.02	NA	NA	NA
	2 wt% FCLS	34.76	24.98	38.95	NA	NA	NA
	0.2 wt% SMT	20.05	12.95	24.41	NA	NA	NA
	0.5 wt% SMT	25.03	13.98	24.80	NA	NA	NA
	1.5 wt% SMT	34.67	19.93	37.92	NA	NA	NA
	2 wt% SMT	37.81	21.01	45.91	NA	NA	NA
	0.2 wt% SHR	17.82	14.97	18.00	NA	NA	NA
	0.5 wt% SHR	19.89	14.84	18.95	NA	NA	NA
	0.8 wt% SHR	24.93	15.97	19.98	NA	NA	NA
	1 wt% SHR	30.80	17.00	23.46	NA	NA	NA
	1.2 wt% SHR	33.76	16.96	23.62	NA	NA	NA
	1.5 wt% SHR	37.78	18.98	24.72	NA	NA	NA
	2 wt% SHR	44.70	20.02	24.88	NA	NA	NA
	1.5 wt% SMT	34.67	19.93	37.92	NA	NA	NA
	2 wt% SMT	37.81	21.01	45.91	NA	NA	NA

Author	System	AV (mPa.S)	PV (mPa.S)	YP (Pa)	Gel (Pa)	Fl (ml)	YP/PV (Pa/mPa.S)
Wang et al. [8]	0.8 wt% KL[(a=−10)]	40.19	NA	0.77	NA	NA	NA
	0.8 wt% KL[(a=−7)]	38.49	NA	0.78	NA	NA	NA

(continued)

Table 6.1 (continued)

Author	System	AV (mPa.S)	PV (mPa.S)	YP (Pa)	Gel (Pa)	Fl (ml)	YP/PV (Pa/mPa.S)
	0.8 wt% KL$^{(a=-4)}$	36.98	NA	0.75	NA	NA	NA
	0.8 wt% KL$^{(a=-1)}$	36.04	NA	0.76	NA	NA	NA
	0.8 wt% KL$^{(a=2)}$	34.53	NA	0.50	NA	NA	NA
	0.8 wt% KL$^{(a=5)}$	32.08	NA	0.54	NA	NA	NA
	0.8 wt% KL$^{(a=8)}$	31.61	NA	0.50	NA	NA	NA
	0.8 wt% KL$^{(a=11)}$	30.57	NA	0.78	NA	NA	NA
	0.4 wt% PAM$^{(a=-10)}$	6.05	NA	0.49	NA	NA	NA
	0.4 wt% PAM$^{(a=-7)}$	5.39	NA	0.24	NA	NA	NA
	0.4 wt% PAM$^{(a=-4)}$	4.55	NA	0.26	NA	NA	NA
	0.4 wt% PAM$^{(a=-1)}$	3.98	NA	0.29	NA	NA	NA
	0.4 wt% PAM$^{(a=2)}$	4.36	NA	0.25	NA	NA	NA
	0.4 wt% PAM$^{(a=5)}$	3.79	NA	0.25	NA	NA	NA
	0.4 wt% PAM$^{(a=8)}$	3.04	NA	0.05	NA	NA	NA
	0.4 wt% PAM$^{(a=11)}$	2.85	NA	0.08	NA	NA	NA
	0.8 wt% XG$^{(a=-10)}$	36.70	NA	5.13	NA	NA	NA
	0.8 wt% XG$^{(a=-7)}$	35.19	NA	4.60	NA	NA	NA
	0.8 wt% XG$^{(a=-4)}$	33.12	NA	4.59	NA	NA	NA
	0.8 wt% XG$^{(a=-1)}$	32.65	NA	4.51	NA	NA	NA
	0.8 wt% XG$^{(a=2)}$	30.95	NA	4.42	NA	NA	NA
	0.8 wt% XG$^{(a=5)}$	30.57	NA	4.07	NA	NA	NA
	0.8 wt% XG$^{(a=8)}$	29.72	NA	3.79	NA	NA	NA

(continued)

Table 6.1 (continued)

Author	System	AV (mPa.S)	PV (mPa.S)	YP (Pa)	Gel (Pa)	Fl (ml)	YP/PV (Pa/mPa.S)
	0.8 wt% XG (a=11)	27.93	NA	3.57	NA	NA	NA
	0.8% KL + 0.6% GC (a=−10)	101.45	NA	5.64	NA	NA	NA
	0.8% KL + 0.6% GC (a=−7)	93.19	NA	5.16	NA	NA	NA
	0.8% KL + 0.6% GC (a=−4)	89.40	NA	5.43	NA	NA	NA
	0.8% KL + 0.6% GC (a=−1)	85.61		4.61	NA	NA	NA

Author	System	AV (mPa.S)	PV (mPa.S)	YP (Pa)	Gel (Pa)	Fl (ml)	YP/PV (Pa/mPa.S)
Wang et al. [8]	0.8% KL + 0.6% GC (a=2)	81.35	NA	4.15	NA	NA	NA
	0.8% KL + 0.6% GC (a=5)	77.09	NA	3.57	NA	NA	NA
	0.8% KL + 0.6% GC (a=8)	72.59	NA	3.62	NA	NA	NA
	0.8% KL + 0.6% GC (a=11)	72.33	NA	3.62	NA	NA	NA
	0.4% PAM + 0.6% GC (a=−10)	16.27	NA	0.27	NA	NA	NA
	0.4% PAM + 0.6% GC (a=−7)	15.54	NA	0.31	NA	NA	NA
	0.4% PAM + 0.6% GC (a=−4)	14.10	NA	0.34	NA	NA	NA
	0.4% PAM + 0.6% GC (a=−1)	12.67	NA	0.29	NA	NA	NA
	0.4% PAM + 0.6% GC (a=2)	12.17	NA	0.14	NA	NA	NA
	0.4% PAM + 0.6% GC (a=5)	11.67	NA	0.14	NA	NA	NA
	0.4% PAM + 0.6% GC (a=8)	10.94	NA	0.14	NA	NA	NA
	0.4% PAM + 0.6% GC (a=11)	9.98	NA	0.14	NA	NA	NA
	0.8% XC + 0.6% GC (a=−10)	60.51	NA	10.32	NA	NA	NA
	0.8% XC + 0.6% GC (a=−7)	55.54	NA	9.50	NA	NA	NA

(continued)

Table 6.1 (continued)

Author	System	AV (mPa.S)	PV (mPa.S)	YP (Pa)	Gel (Pa)	Fl (ml)	YP/PV (Pa/mPa.S)
	0.8% XC + 0.6% GC (a=−4)	52.69	NA	9.77	NA	NA	NA
	0.8% XC + 0.6% GC (a=−1)	51.02	NA	9.77	NA	NA	NA
	0.8% XC + 0.6% GC (a=2)	50.29	NA	9.77	NA	NA	NA
	0.8% XC + 0.6% GC (a=5)	48.38	NA	9.29	NA	NA	NA
	0.8% XC + 0.6% GC (a=8)	48.35	NA	9.26	NA	NA	NA
	0.8% XC + 0.6% GC (a=11)	47.39	NA	9.24	NA	NA	NA
Zhang et al. [9]	BM-18	NA	32	15	NA	4.8	0.47
	10 wt% NaCl	NA	27	22	NA	5.1	0.81
	15 wt% NaCl	NA	30	23	NA	4.9	0.77
	10 wt% EG	NA	37	11	NA	4.6	0.30
	15 wt% EG	NA	41	7.5	NA	4.4	0.18
Nikolaev et al. [2]	a = −4.0	NA	20.07	9.56	2.87/3.62	5.51	0.48
	−2.0	NA	19.66	9.42	2.89/3.7	5.05	0.48
	0.0	NA	19.26	9.23	2.32/3.24	5.02	0.48
	2.0	NA	19.06	8.93	2.47/3.26	4.83	0.47
	4.0	NA	18.66	8.73	2.37/3.13	4.64	0.47

Author	System	AV (mPa.S)	PV (mPa.S)	YP (Pa)	Gel (Pa)	Fl (ml)	YP/PV (Pa/mPa.S)
Nikolaev et al. [2]	0.4 wt% PVP K-90	NA	13.5	5.2	2/2.5	7.5	0.39
	0.5 wt% PVP K-90	NA	14	5.25	2.5/3	7.2	0.38
	0.6 wt% PVP K-90	NA	14	6.25	2/3	6.8	0.45

(continued)

Table 6.1 (continued)

Author	System	AV (mPa.S)	PV (mPa.S)	YP (Pa)	Gel (Pa)	Fl (ml)	YP/PV (Pa/mPa.S)
	0.7 wt% PVP K-90	NA	16	6.78	2,5/3	6.2	0.42
	0.8 wt% PVP K-90	NA	17.5	7.66	2,5/3	5.5	0.44
	0.9 wt% PVP K-90	NA	18.5	8.82	3/3,5	5.3	0.48
	1 wt% PVP K-90	NA	19.5	9.36	3/3,5	5	0.48
	1.1 wt% PVP K-90	NA	22	8.86	3/4	5	0.40
	1.2 wt% PVP K-90	NA	23.5	9.15	3,5/4,5	5	0.39
Jiang et al. [10]	a = -8	NA	18	8.5	2/2,5	5.5	0.47
	-4	NA	20	8.9	2/2,5	5.5	0.45
	0	NA	21	9.2	2,5/3	5.5	0.44
	8	NA	22	9.7	2,5/3	5.7	0.44
	15	NA	25	10.2	2,5/3	5.8	0.41
	0.3% MgCl2 + 0.1% CaCl2	NA	18.5	9.2	2,5/3	5.5	0.50
	0.51%MgCl2 + 0.18% CaCl2	NA	20	9	3/3,5	5.8	0.45
	0.8% MgCl2 + 0.4% CaCl2	NA	21.5	9	3/3,5	5.8	0.42
	0.4 wt% PVP K-90	NA	12	5.1	1,5/2	11	0.43
	0.5 wt% PVP K-90	NA	13	5.1	1,5/2	10.8	0.39
	0.6 wt% PVP K-90	NA	14.5	6.6	1/2	10.4	0.46
	0.7 wt% PVP K-90	NA	16.8	3.6	1,5/2,5	9.5	0.21
	0.8 wt% PVP K-90	NA	18	4.6	1,5/2	7.8	0.26
	0.9 wt% PVP K-90	NA	18.5	8.7	2/2,5	6.5	0.47
	1 wt% PVP K-90	NA	20	9.2	2/2,5	5.5	0.46

(continued)

Table 6.1 (continued)

Author	System	AV (mPa.S)	PV (mPa.S)	YP (Pa)	Gel (Pa)	Fl (ml)	YP/PV (Pa/mPa.S)
	1.1 wt% PVP K-90	NA	22.5	8.7	2/3	5.5	0.39
	1.2 wt% PVP K-90	NA	23	9.7	2.5/4.5	5.5	0.42
Wang et al. [11]	0.1 wt%Modified starch[a=5]	2	1.9	NA	NA	NA	NA
	0.2 wt%Modified starch[a=5]	2.5	2.4	NA	NA	NA	NA
	0.3 wt%Modified starch[a=5]	2.85	2.8	NA	NA	NA	NA
	0.4 wt%Modified starch[a=5]	2.5	2.6	NA	NA	NA	NA
	0.5 wt%Modified starch[a=5]	2.15	2.3	NA	NA	NA	NA
	0.1 wt%Modified starch[a=10]	1.75	1.8	NA	NA	NA	NA
	0.2 wt Modified starch[a=10]	1.6	1.7	NA	NA	NA	NA
	0.3 wt%Modified starch[a=10]	1.5	1.6		NA		NA

Author	System	AV (mPa.S)	PV (mPa.S)	YP (Pa)	Gel (Pa)	Fl (ml)	YP/PV (Pa/mPa.S)
Wang et al. [11]	0.4 wt Modified starch[a=10]	2.5	2.4	NA	NA	NA	NA
	0.5 wt Modified starch[a=10]	2.25	2.3	NA	NA	NA	NA
	0.1 wt%Modified starch[a=15]	1.6	1.7	NA	NA	NA	NA
	0.2 wt%Modified starch[a=15]	1.5	1.6	NA	NA	NA	NA
	0.3 wt Modified starch[a=15]	2.5	2.4	NA	NA	NA	NA
	0.4 wt Modified starch[a=15]	2.25	2.3	NA	NA	NA	NA
	0.5 wt Modified starch[a=15]	2.05	2.2	NA	NA	NA	NA
	0.1 wt Modified starch[a=20]	1.5	1.6	NA	NA	NA	NA
	0.2 wt Modified starch[a=20]	2.5	2.4	NA	NA	NA	NA

(continued)

Table 6.1 (continued)

Author	System	AV (mPa.S)	PV (mPa.S)	YP (Pa)	Gel (Pa)	Fl (ml)	YP/PV (Pa/mPa.S)
	0.3 wt Modified starch[a=20]	2.25	2.3	NA	NA	NA	NA
	0.4 wt Modified starch[a=20]	2.05	2.2	NA	NA	NA	NA
	0.5 wt Modified starch[a=20]	1.95	2.1	NA	NA	NA	NA
	0.1 wt% CMC[a=5]	7.45	5.9	NA	NA	NA	NA
	0.2 wt% CMC [a=5]	14.65	10.8	NA	NA	NA	NA
	0.3 wt% CMC [a=5]	19.15	13.5	NA	NA	NA	NA
	0.4 wt% CMC [a=5]	23.55	16.2	NA	NA	NA	NA
	0.5 wt% CMC [a=5]	31.8	20.9	NA	NA	NA	NA
	0.1 wt% CMC [a=10]	6.8	5.5	NA	NA	NA	NA
	0.2 wt% CMC [a=10]	13.55	10.2	NA	NA	NA	NA
	0.3 wt% CMC [a=10]	17.65	12.8	NA	NA	NA	NA
	0.4 wt% CMC [a=10]	22.05	16	NA	NA	NA	NA
	0.5 wt% CMC [a=10]	29.25	20	NA	NA	NA	NA
	0.1 wt% CMC [a=15]	6.1	5.1	NA	NA	NA	NA
	0.2 wt% CMC [a=15]	12.35	9.6	NA	NA	NA	NA
	0.3 wt% CMC [a=15]	16.25	12.1	NA	NA	NA	NA
	0.4 wt% CMC [a=15]	20.25	14.9	NA	NA	NA	NA
	0.5 wt% CMC [a=15]	27.15	18.9	NA	NA	NA	NA
	0.1 wt% CMC [a=20]	5.7	4.8	NA	NA	NA	NA
	0.2 wt% CMC [a=20]	11.2	8.6	NA	NA	NA	NA
	0.3 wt% CMC [a=20]	14.95	11.4	NA	NA	NA	NA

(continued)

Table 6.1 (continued)

Author	System	AV (mPa.S)	PV (mPa.S)	YP (Pa)	Gel (Pa)	Fl (ml)	YP/PV (Pa/mPa.S)
	0.4 wt% CMC [a=20]	18.35	13.8	NA	NA	NA	NA
	0.5 wt% CMC [a=20]	24.9	18	NA	NA	NA	NA
	0.1 wt% XG [a=5]	6.45	4.4	NA	NA	NA	NA

Author	System	AV (mPa.S)	PV (mPa.S)	YP (Pa)	Gel (Pa)	Fl (ml)	YP/PV (Pa/mPa.S)
Wang et al. [11]	0.2 wt% XG [a=5]	6	4.1	NA	NA	NA	NA
	0.3 wt% XG [a=5]	5.5	3.9	NA	NA	NA	NA
	0.4 wt% XG [a=5]	5.2	3.4	NA	NA	NA	NA
	0.5 wt% XG [a=5]	10	5.3	NA	NA	NA	NA
	0.1 wt% XG [a=10]	6	4.1	NA	NA	NA	NA
	0.2 wt% XG [a=10]	5.5	3.9	NA	NA	NA	NA
	0.3 wt% XG [a=10]	5.2	3.4	NA	NA	NA	NA
	0.4 wt% XG [a=10]	10	5.3	NA	NA	NA	NA
	0.5 wt% XG [a=10]	9.3	4.3	NA	NA	NA	NA
	0.1 wt% XG [a=15]	5.5	3.9	NA	NA	NA	NA
	0.2 wt% XG [a=15]	5.2	3.4	NA	NA	NA	NA
	0.3 wt% XG [a=15]	10	5.3	NA	NA	NA	NA
	0.4 wt% XG [a=15]	9.3	4.3	NA	NA	NA	NA
	0.5 wt% XG [a=15]	8.6	4	NA	NA	NA	NA
	0.1 wt% XG [a=20]	5.2	3.4	NA	NA	NA	NA
	0.2 wt% XG [a=20]	10	5.3	NA	NA	NA	NA

(continued)

Table 6.1 (continued)

Author	System	AV (mPa.S)	PV (mPa.S)	YP (Pa)	Gel (Pa)	Fl (ml)	YP/PV (Pa/mPa.S)
	0.3 wt% XG (a=20)	9.3	4.3	NA	NA	NA	NA
	0.4 wt% XG (a=20)	8.6	4	NA	NA	NA	NA
	0.5 wt% XG (a=20)	8.15	3.7	NA	NA	NA	NA
Saikia and Mahto [12]	BM-21[a=2]	16	29.5	NA	NA	NA	NA
	BM-21[a=10]	14	26	NA	NA	NA	NA
	BM-21[a=20]	13	22	NA	NA	NA	NA
	BM-21[a=25]	11	20	NA	NA	NA	NA
	0.1 wt% PVP[a=2]	16	29.5	NA	NA	NA	NA
	0.5 wt% PVP [a=2]	13	22	NA	NA	NA	NA
	1 wt% PVP [a=2]	13	23.5	NA	NA	NA	NA
	0.1 wt% PVP[a=10]	14	26	NA	NA	NA	NA
	0.5 wt% PVP [a=10]	16	29.5	NA	NA	NA	NA
	1 wt% PVP [a=10]	13	22	NA	NA	NA	NA
	0.1 wt% PVP[a=20]	13	22.5	NA	NA	NA	NA
	0.5 wt% PVP [a=20]	14	26	NA	NA	NA	NA
	1 wt% PVP [a=20]	16	29.5	NA	NA	NA	NA

Author	System	AV (mPa.S)	PV (mPa.S)	YP (Pa)	Gel (Pa)	Fl (ml)	YP/PV (Pa/mPa.S)
Saikia and Mahto [12]	0.1 wt% PVP[a=25]	11	20.5	NA	NA	NA	NA
	0.5 wt% PVP [a=25]	13	22.5	NA	NA	NA	NA
	1 wt% PVP [a=25]	14	26	NA	NA	NA	NA

(continued)

Table 6.1 (continued)

Author	System	AV (mPa.S)	PV (mPa.S)	YP (Pa)	Gel (Pa)	Fl (ml)	YP/PV (Pa/mPa.S)
	0.1 wt% Lipase [a=2]	17	30	NA	NA	NA	NA
	0.5 wt% Lipase [a=2]	12	21	NA	NA	NA	NA
	1 wt% Lipase [a=2]	13	22.5	NA	NA	NA	NA
	0.1 wt% Lipase [a=10]	15	27	NA	NA	NA	NA
	0.5 wt% Lipase [a=10]	17	30	NA	NA	NA	NA
	1 wt% Lipase [a=10]	12	21	NA	NA	NA	NA
	0.1 wt% Lipase [a=20]	13	22	NA	NA	NA	NA
	0.5 wt% Lipase [a=20]	15	27	NA	NA	NA	NA
	1 wt% Lipase [a=20]	17	30	NA	NA	NA	NA
	0.1 wt% Lipase [a=25]	12	19.5	NA	NA	NA	NA
	0.5 wt% Lipase [a=25]	13	22	NA	NA	NA	NA
	1 wt% Lipase [a=25]	15	27	NA	NA	NA	NA
Saikia et al. [1]	BM-21[a=1]	18	31.5	12.0	NA	NA	0.67
	BM-21[a=10]	17	28	10.5	NA	NA	0.62
	BM-21[a=20]	16	24	7.7	NA	NA	0.48
	BM-21[a=25]	14	22	7.7	NA	NA	0.55
	0.1 wt% EG[a=1]	18	31.5	25	NA	NA	12.0
	0.3 wt% EG [a=1]	15	22.5	15	7/13	10	7.2
	0.5 wt% EG [a=1]	19	32	24	NA	NA	11.5
	0.7 wt% EG [a=1]	15	22.5	15	NA	NA	7.2
	1 wt% EG[a=1]	17	24.5	15	NA	NA	7.2

(continued)

Table 6.1 (continued)

Author	System	AV (mPa.S)	PV (mPa.S)	YP (Pa)	Gel (Pa)	FI (ml)	YP/PV (Pa/mPa.S)
	3 wt% EG [a=1]	18	28.5	21	NA	NA	10.1
	5 wt% EG [a=1]	19	32	24	NA	NA	11.5
	0.1 wt% EG [a=10]	17	28	22	NA	NA	0.62
	0.3 wt% EG [a=10]	18	31.5	25	NA	NA	0.67
	0.5 wt% EG [a=10]	15	22.5	15	NA	NA	0.48
	0.7 wt% EG [a=10]	17	24.5	15	NA	NA	0.42
	1 wt% EG [a=10]	18	29	22	NA	NA	0.56
	3 wt% EG [a=10]	18	29	22	NA	NA	10.5
	5 wt% EG [a=01]	19	33	25	NA	NA	12.0
	0.1 wt% EG [a=20]	16	24	16	NA	NA	7.7

Author	System	AV (mPa.S)	PV (mPa.S)	YP (Pa)	Gel (Pa)	FI (ml)	YP/PV (Pa/mPa.S)
Saikia et al. [1]	0.3 wt% EG [a=20]	17	28	22	NA	NA	10.5
	0.5 wt% EG [a=20]	18	31.5	25	NA	NA	12.0
	0.7 wt% EG [a=20]	15	22.5	15	NA	NA	7.2
	1 wt% EG [a=20]	17	24.5	15	NA	NA	7.2
	3 wt% EG [a=20]	18	28.5	21	NA	NA	10.1
	5 wt% EG [a=20]	19	32	24	NA	NA	11.5
	0.1 wt% EG [a=25]	14	22	16	NA	NA	7.7
	0.3 wt% EG [a=25]	16	24	16	NA	NA	7.7
	0.5 wt% EG [a=25]	17	28	22	NA	NA	10.5

(continued)

Table 6.1 (continued)

Author	System	AV (mPa.S)	PV (mPa.S)	YP (Pa)	Gel (Pa)	FI (ml)	YP/PV (Pa/mPa.S)
	0.7 wt% EG [a=25]	18	31.5	25	NA	NA	12.0
	1 wt% EG [a=25]	15	22.5	15	NA	NA	7.2
	3 wt% EG [a=25]	17	24.5	15	NA	NA	7.2
	5 wt% EG [a=25]	18	28.5	21	NA	NA	10.1
	0.1 wt% Carbon Dots [a=1]	17	31.5	27	NA	NA	12.9
	0.3 wt% Carbon Dots [a=1]	13	22.5	19	7/13	10	9.1
	0.5 wt% Carbon Dots [a=1]	15	24.5	19	NA	NA	9.1
	0.7 wt% Carbon Dots [a=1]	16	28.5	25	NA	NA	12.0
	1 wt% Carbon Dots [a=1]	17	32	28	NA	NA	13.4
	0.1 wt% Carbon Dots [a=10]	16	28	24	NA	NA	11.5
	0.3 wt% Carbon Dots [a=10]	17	31.5	27	NA	NA	12.9
	0.5 wt% Carbon Dots [a=10]	13	22.5	19	NA	NA	9.1
	0.7 wt% Carbon Dots [a=10]	15	24.5	19	NA	NA	9.1
	1 wt% Carbon Dots [a=10]	16	28.5	25	NA	NA	12.0
	0.1 wt% Carbon Dots [a=20]	15	24	18	NA	NA	8.6
	0.3 wt% Carbon Dots [a=20]	16	28	24	NA	NA	11.5
	0.5 wt% Carbon Dots [a=20]	17	31.5	27	NA	NA	12.9
	0.7 wt% Carbon Dots [a=20]	13	22.5	19	NA	NA	9.1
	1 wt% Carbon Dots [a=20]	15	24.5	19	NA	NA	9.1
	0.1 wt% Carbon Dots [a=25]	13	22	18	NA	NA	8.6
	0.3 wt% Carbon Dots [a=25]	15	24	18	NA	NA	8.6

(continued)

Table 6.1 (continued)

Author	System	AV (mPa.S)	PV (mPa.S)	YP (Pa)	Gel (Pa)	FI (ml)	YP/PV (Pa/mPa.S)
	0.5 wt% Carbon Dots [a=25]	16	28	24	NA	NA	11.5
	0.7 wt% Carbon Dots [a=25]	17	31.5	27	NA	NA	12.9
	1 wt% Carbon Dots [a=25]	13	22.5	19	NA	NA	9.1

Author	System	AV (mPa S)	PV (mPa S)	YP (Pa)	Gel (Pa)	FI (ml)	YP/PV (Pa/mPa S)
Saikia et al. [1]	BM-22	20	19.5	NA	NA	NA	0.98
	0.3 wt%PVP	22.5	21	NA	NA	NA	0.93
	0.5 wt%PVP	23.5	22.5	NA	NA	NA	0.96
	1 wt%PVP	27	26	NA	NA	NA	0.96
	10 wt%NaOH	20	19.5	NA	NA	NA	0.98
	10 wt%NaOH + 0.5 wt%PVP	22	21.5	NA	NA	NA	0.98
	10 wt%NaOH + 1 wt%PVP	24	25	NA	NA	NA	1.04
	10 wt%NaOH + 1.5 wt%PVP	26	26	NA	NA	NA	1.00
	10 wt%NaOH + 2 wt%PVP	30.5	31	NA	NA	NA	1.02
Hale and Dewan [13]	84 mlH$_2$O + 0.48XG + 21gNaCl + 253glycerol [a=23.8]	THTM	NA	NA	16/21	NA	NA
	105 mlH$_2$O + 0.6XG + 26gNaCl + 222glycerol [a=23.8]	146	116	28.73	14/48	NA	0.25
	125 mlH$_2$O + 0.72XG + 31gNaCl + 192glycerol [a=23.8]	83	33	47.40	12/25	NA	1.44
	146 mlH$_2$O + 0.83XG + 36gNaCl + 162glycerol [a=23.8]	71	57	13.41	10/13	NA	0.24

(continued)

Table 6.1 (continued)

Author	System	AV (mPa S)	PV (mPa S)	YP (Pa)	Gel (Pa)	Fl (ml)	YP/PV (Pa/mPa S)
	166 mlH$_2$O + 0.95XG + 41gNaCl + 132glycerol[(a=23.8)]	61	50	10.53	11/12	NA	0.21
	186 mlH$_2$O + 1.06XG + 46gNaCl + 102glycerol[(a=23.8)]	51	35	15.32	11/13	NA	0.44
	206 mlH$_2$O + 1.18XG + 51gNaCl + 73glycerol[(a=23.8)]	46	28	16.76	11/15	NA	0.60
	226 mlH$_2$O + 1.29XG + 56gNaCl + 43glycerol[(a=23.8)]	44	27	16.28	12/16	NA	0.60
	245 mlH$_2$O + 1.4XG + 61gNaCl + 14glycerol[(a=23.8)]	43	29	12.93	11/15	NA	0.45
	84 mlH$_2$O + 0.48XG + 21gNaCl + 253glycerol[(a=6.7)]	THTM	NA	NA	14/70	NA	
	105 mlH$_2$O + 0.6XG + 26gNaCl + 222glycerol[(a=6.7)]	244	216	26.81	14/42	NA	0.12
	125 mlH$_2$O + 0.72XG + 31gNaCl + 192glycerol[(a=6.7)]	THTM	NA	NA	12/24	NA	
	146 mlH$_2$O + 0.83XG + 36gNaCl + 162glycerol[(a=6.7)]	126	107	18.19	12/15	NA	0.17
	166 mlH$_2$O + 0.95XG + 41gNaCl + 132glycerol[(a=6.7)]	92	73	18.19	12/14	NA	0.25
	186 mlH$_2$O + 1.06XG + 46gNaCl + 102glycerol[(a=6.7)]	73	53	18.67	13/16	NA	0.35
	206 mlH$_2$O + 1.18XG + 51gNaCl + 73glycerol[(a=6.7)]	65	45	18.67	13/16	NA	0.41

(continued)

Table 6.1 (continued)

Author	System	AV (mPa S)	PV (mPa S)	YP (Pa)	Gel (Pa)	Fl (ml)	YP/PV (Pa/mPa S)
	226 mlH$_2$O + 1.29XG + 56NaCl + 43glycerol[a=6.7]	58	38	18.67	14/17	NA	0.49
	245 mlH$_2$O + 1.4XG + 61gNaCl + 14glycerol[a=6.7]	55	36	17.72	13/16	NA	0.49
	125 mlH$_2$O + 0.72XG + 31gNaCl + 192glycerol[a=6.7]	THTM	NA	NA	12/24	NA	NA
	84 mlH$_2$O + 0.48XG + 21gNaCl + 253glycerol[a=-15]	THTM	NA	NA	24/54	NA	NA
	105 mlH$_2$O + 0.6XG + 26gNaCl + 222glycerol[a=-15]	THTM	NA	NA	22/30	NA	NA
	125 mlH$_2$O + 0.72XG + 31gNaCl + 192glycerol[a=-15]	THTM	NA	NA	20/24	NA	NA
	146 mlH$_2$O + 0.83XG + 36gNaCl + 162glycerol[a=-15]	THTM	NA	NA	18/22	NA	NA
	166 mlH$_2$O + 0.95XG + 41gNaCl + 132glycerol[a=-15]	174	148	24.90	16/20	NA	0.17

Author	System	AV (mPa.S)	PV (mPa.S)	YP (Pa)	Gel (Pa)	Fl (ml)	YP/PV (Pa/mPa.S)
Hale and Dewan [13]	186 mlH$_2$O + 1.06XG + 46gNaCl + 102glycerol[a=-15]	162	140	20.59	15/19	NA	0.15
	206 mlH$_2$O + 1.18XG + 51gNaCl + 73glycerol[a=-15]	122	97	23.46	16/19	NA	0.24
	226 mlH$_2$O + 1.29XG + 56gNaCl + 43glycerol[a=-15]	98	74	22.50	16/19	NA	0.30

(continued)

Table 6.1 (continued)

Author	System	AV (mPa.S)	PV (mPa.S)	YP (Pa)	Gel (Pa)	FI (ml)	YP/PV (Pa/mPa.S)
	245 mlH$_2$O + 1.4XG + 61gNaCl + 14glycerol[a=-15]	86	71	19.15	15/17	NA	0.27
	PFFM	70	45	23.94	10/17	11.6	0.53
	PFFM + 10 wt%glycerol	70	44	24.90	11/18	11	0.57
	PFFM + 20 wt%glycerol	69.5	45	23.46	12/17	10	0.52
	PFFM + 30 wt%glycerol	69	45	22.98	9/15	9.5	0.51
	PFFM + 40 wt%glycerol	67	43	22.98	8/13	8.1	0.53
	BM-23	70	45	23.94	10/17	11.6	0.53
	10 vv.%Pill	71	47	22.98	9/16	11	0.49
	20 vv.%Pill	68.5	40	27.29	8/14	10	0.68
	30 vv.%Pill	64.5	43	20.59	9/12	9.4	0.48
	40 vv.%Pill	62.5	41	20.59	8/11	9.6	0.50
	10 wt%glycerol	67.5	46	20.59	7/12	9.2	0.45
	23 wt%glycerol	65	47	17.24	5/8	7.8	0.37
	46 wt%glycerol	64	50	13.41	3/4	6	0.27
	MFMP	38	32	5.75	3/13	3.2	0.18
	10 vv.%Pill	38	31	6.70	3/10	3.8	0.22
	20 vv.%Pill	36.5	29	7.18	3/8	3.8	0.25
	35 vv.%Pill	39.5	30	9.10	3/7	3.8	0.30
	50 vv.%Pill	35.5	30	5.27	3/6	4.2	0.18
Grigg and Lynes [14]	BM-24 [a=121]	41.5	34	6.7	4/5	NA	0.20
	19.22-wt% CaCl2 Brine[a=121]	55.4	45	10	4/6	NA	0.22

(continued)

Table 6.1 (continued)

Author	System	AV (mPa.S)	PV (mPa.S)	YP (Pa)	Gel (Pa)	Fl (ml)	YP/PV (Pa/mPa.S)
Yousif and Young [15]	Basic spud mud[a=49]	69	13	53.6	22/37	7.4	4.13
	Spud mud + 4000 ppm lecithin[a=49]	39	20	18.2	9/22	7	0.91
	Spud mud + LTL[a=49]	41	24	16.3	8/22	6.6	0.68

BM-8(2 wt% Bentonite + 0.25 wt% FA-367 + 0.5 wt%PAC-LV + 2 wt%SMP + 5 wt%KCl + 10 wt%NaCl); BM-9(Artificial sea water + 3 wt% Bentonite + 3 wt% Barite + 0.3 wt% CMC); BM-10(Pre-hydrated bentonite mud); BM-11(Simple polymer mud); BM-12(KCL/PHPA polymer mud); BM-13(Advanced polymer mud); BM-14(Polyacrylate-based fluid); BM-15(Advanced polymer mud); BM-16(Water + natural calcium bentonite (5%) + sodium carbonate (6%) + 10% NaCl + 5% KCl + 1 ‰ NaOH); MFV(Mash Funnel viscosity/s)

BM-17(1L H2O + 15% NaCl + 0.5% NaOH)

BM-18(4% bentonite + 0.3% ZCJ-1 + 0.1% ZCJ-2 + 0.5% JYJ-1 + 3% JYJ-2 + 3% SD-102); BM-19(water + 6% polyethylene glycol + 4% propylene glycol + 3%bentonite + 3%NaCl + 5% Na-CMC + 2.5% SMP-2); BM-20(3% sodium bentonite, 1% LV-PAC, 4% SMP-2, 10% polyethylene glycol, 20% NaCl, and 0.5% NaOH)

BM-21(Distilled water, THF, 5 KCl + 0.4 CMC + 0.4 XG + 0.4 PAC)

BM-22(4 wt% Bentonite + 0.1 wt% XG + 0.4 wt%PC + 0.4 wt% Starch + 4 wt%KCl + 0.3 wt%NaOH)

PFFM (Pill formulation with field mud-high salt polymer mud); BM-23(2lb/bbl of XC polymer in the water phase, 20% salt, and 50% glycerol); MFMP (Mexico filed mud with pill) –0.72Ib/bbl of XC in 20% NaCl); BM-24(0.709 gal LVT oil + 0.19 lbm Invermul + 0.19 lbm Lime + 0.18 gal water + 0.14lbm Gelton + 0.05lbm EZ Mul + 0.71lbmRev-Dust + 0.29lbm Duraton + 2.86lbm Barite); LTL (4000 ppm lecithin + 1 wt% Torq-Trim II lubricant) for drilling fluids in drilling gas hydrate sediments, it is encouraging to study new ILs to develop effective and efficient drilling mud system for drilling gas hydrate sediments.

References

1. Saikia T, Mahto V, Kumar A (2017) Quantum dots: a new approach in thermodynamic inhibitor for the drilling of gas hydrate bearing formation. J Ind Eng Chem 52:89–98
2. Nikolaev NI, Liu T, Wang Z, Jiang G, Sun J, Zheng M, Wang Y (2014) The experimental study on a new type low temperature water-based composite alcohol drilling fluid. Procedia Eng 73:276–282
3. Zhao X, Qiu Z, Zhao C, Xu J, Zhang Y (2019) Inhibitory effect of water-based drilling fluid on methane hydrate dissociation. Chem Eng Sci 199:113–122
4. Srungavarapu M, Patidar KK, Pathak AK, Mandal A (2018) Performance studies of water-based drilling fluid for drilling through hydrate bearing sediments. Appl Clay Sci 152:211–220
5. Fereidounpour A, Vatani A (2014) An investigation of interaction of drilling fluids with gas hydrates in drilling hydrate bearing sediments. J Nat Gas Sci Eng 20:422–427
6. Fereidounpour A, Vatani A (2015) Designing a Polyacrylate drilling fluid system to improve wellbore stability in hydrate bearing sediments. J Nat Gas Sci Eng 26:921–926
7. Chen L, Wang S, Ye C (2014) Effect of gas hydrate drilling fluids using low solid phase mud system in plateau permafrost. Procedia Eng 73:318–325
8. Wang S, Yuan C, Zhang C, Chen L, Liu J (2017) Rheological properties with temperature response characteristics and a mechanism of solid-free polymer drilling fluid at low temperatures. Appl Sci 7:18
9. Zhang H, Cheng Y, Shi J, Li L, Li M, Han X, Yan C (2017) Experimental study of water-based drilling fluid disturbance on natural gas hydrate-bearing sediments. J Nat Gas Sci Eng 47:1–10
10. Jiang G, Liu T, Ning F, Tu Y, Zhang L, Yu Y, Kuang L (2011) Polyethylene glycol drilling fluid for drilling in marine gas hydrates-bearing sediments: an experimental study. Energies 4:140–150
11. Wang R, Sun H, Shi X, Xu X, Zhang L, Zhang Z (2019) Fundamental investigation of the effects of modified starch, carboxymethylcellulose sodium, and xanthan gum on hydrate formation under different driving forces. Energies 12:2026
12. Saikia T, Mahto V (2016) Experimental investigations of clathrate hydrate inhibition in water based drilling fluid using green inhibitor. J Pet Sci Eng 147:647–653
13. Hale AH, Dewan AKR (1990) Inhibition of gas hydrates in deepwater drilling. SPE Drill Eng 5(109–115):18638
14. Grigg RB, Lynes GL (1992) Oil-based drilling mud as a gas-hydrates inhibitor. SPE Drill Eng 7:32–38
15. Yousif MH, Young DB (1993) Simple correlation to predict the hydrate point suppression in drilling fluids. Drill Conf—Proc 287–294
16. Zhao X, Qiu Z, Zhou G, Huang W (2015) Synergism of thermodynamic hydrate inhibitors on the performance of poly (vinyl pyrrolidone) in deepwater drilling fluid. J Nat Gas Sci Eng 23:47–54
17. Saikia T, Mahto V (2018) Experimental investigations and optimizations of rheological behavior of drilling fluids using RSM and CCD for gas hydrate-bearing formation. Arab J Sci Eng 43:1–14
18. Jiang G, Ning F, Zhang L, Tu Y (2011) Effect of agents on hydrate formation and low-temperature rheology of polyalcohol drilling fluid. J Earth Sci 22:652–657
19. Luo Z, Pei J, Wang L, Yu P, Chen Z (2017) Influence of an ionic liquid on rheological and filtration properties of water-based drilling fluids at high temperatures. Appl Clay Sci 136:96–102

20. Ofei TN, Bavoh CB, Rashidi AB (2017) Insight into ionic liquid as potential drilling mud additive for high temperature wells. J Mol Liq 242:931–939
21. Saikia T, Mahto V (2016) Evaluation of 1-decyl-3-methylimidazolium tetrafluoroborate as clathrate hydrate crystal inhibitor in drilling fluid. J Nat Gas Sci Eng 36:906–915

Chapter 7
Practical Application of Drilling Mud in Hydrate Related Drilling Operations

Gas hydrate formation in drilling muds and the decomposition of in-situ hydrates from the hydrate zones are two serious problems that must be considered for safe operation and production in oil development projects. The problems caused by gas hydrates become more and more frequent as the exploration and production activities are taking place in remote offshore and deep environments. Also, dealing with hydrate production and drilling hydrate sediments demands innovative drilling mud systems for drilling success in hydrate sediments.

7.1 Hydrate Encounter and Mitigation in Offshore Operation

The encounter of hydrate formation in drilling mud was first notices in the late 90s, which followed an immediate research attention owing to the success of hydrate mitigation in pipelines [1]. For instance, between 1983 and 1991 about 25 cases of hydrate formation in drilling mud systems were reported from several offshore oil and gas drilling operations [2]. Since then several drilling mud inhibitive additives to deal with hydrates has been proposed by researchers and industries alike. From a thermodynamic viewpoint, the pressure and temperature conditions surrounding drilling activities in deep waters potentially necessitate hydrate formation, which eventually leads to plugs, if not properly managed. According to Barker and Gomez [3], most offshore hydrate problems occur the subsea equipment which leads to drilling challenges. However, to mitigate these challenges, Hale and Dewan [4] and other researchers at the time proposed the use of chemical inhibitors and drilling parameters optimization to practically prevent hydrate challenges while drilling. The initially proposed additives were based on the additives used for hydrate management in flow assurance, thus, additives such as salts, glycols, alcohol, and their mixtures were adopted. Ouar et al. [5] later claimed that electrolytes and alcohols (methanol) inhibits

© The Author(s), under exclusive license to Springer Nature Switzerland AG 2022
B. Lal et al., *Hydrate Control in Drilling Mud*,
SpringerBriefs in Petroleum Geoscience & Engineering,
https://doi.org/10.1007/978-3-030-94130-7_7

hydrate formation in drilling mud more than the existing conventional inhibitors. The presence of other inhibitors, other than electrolytes and methanol has little hydrate mitigation effect. There are several instances of using kinetic hydrate inhibitors for Deepwater drilling operation as well, however, the thermodynamic inhibitors have proven to be more effective in practical application. Though there has been significant research in literature that introduces new efficient kinetic inhibitors.

7.2 Practical Field Cases

7.2.1 Gulf of Mexico Scenarios

The Gulf of Mexico is the most famous location where drilling operation hydrate-related problems are reported or noticed. In this section, a few such incidents are provided to show practical evidence of the issue. While drilling an offshore well in the Gulf of Mexico located on the US west coast, a hydrate formation occurred and interrupted the drilling operation. The depth of the well-drilled was 350.52 m with a mudline with a temperature of 277.15 K. The hydrate problem was suspected after detecting gas leak while resuming drilling through the casing shoe after casing was completed in an intermediate section. Upon changing the BOP ram sizes, the gas influx increased sporadically with time leading to the closure of the BOPs, while increasing the casing pressure to 9 MPa. The gas was rising up through the cement annulus from the formation at the depth of 2362.2 m, which subsequently lead to the leaking of the wellhead hanger. The leaked gas through the wellhead hanger moving in the freshwater mud at the subsea wellhead (where the temperature condition favors hydrate formation). An attempt the recover the BOP wear bushing failed due to hydrate formation in the mud systems. The inability to retrieve the BOP wear denied access to the leaking casing hanger. Probably due to the hydrate plug formation, the using of drilling mud (primary control method) to balance the pressure at 20.7 MPa also failed. The gas influx was stopped by pumping heavy mud in the formation by perforating the casing around the gas influx location. Eventually, the gas influx was killed the pumping 1.68 gcm^{-3} mud down the drillstring and into the formation at surface pressures up to 21.4 MPa after 7 days. The killing operation confirmed the plugging of the chokeline and the kill line by hydrate at the bottom eight riser joints [3].

Also, while drilling a 944.88 m water in the Gulf of Mexico there was hydrate formation in the drilling mud at a subsea temperature of 277.15 K. While performing a flow check at a drilled depth of 2340.6 m, a kick was detected, hence the pressure of the well was being determined after shutting in for 14 h. To resume drilling, the pressure in the casing was zero while the BOP opens went static. There was a failure to establish circulation using different mud systems after another 14 h. After some time, it was noticed that the choke line and the kill line were plugged by trying to circulate mud up the riser. There were several trials to unplug the kill line and choke line via pressure surges, however, all of such attempts failed. Logging data indicated that there was gas in the drillstring below and above the BOPs in the casing annulus. It was assumed that the gas migration was due to a leak in the wellhead casing hanger. The hydrates and gas in the wellbore were removed by circulating hot mud after perforating through the gas/liquid contact at 121.92 m in the drillstring via the annulus. Three perforations were used to release the gas upon circulating the hot mud. The well was secured and drilling resumed [6].

7.2.2   The Cascade Scenario

The Cascade field is in the USA, where it is believed to has lots of hydrates wells due to its rich gas potentials. From 1994 to 1996 several wells were drilled in the area to develop the field. The field data in the area shows high hydrate risk scatted around the field with a high probability of drilling through hydrates. This section briefly discusses some hydrate-related drilling challenges in the Cascade field. The issue of gas influx in the well begun with the discovery well (Cascade #1). While drilling the Cascade #1, at a depth of 722.7 m, gas was encountered after drilling a hydrate zone. The gas influx was managed via increasing the mud weight and performing. Throughout the drilling operations, all the subsequent hydrate sections were drilling by adjusting the drilling mud weight and circulating the gas out of the well from beneath the surface casing shoe on every trip. After drilling the discovery well, the development well (K-17) spent about 48 h tackling gas kicks. The development well was also disturbed by the gas to the extent that most of the drilling mud was contaminated with the gas, however, just like the development well, increasing the mud weight helped to deal with the issue of hydrates. In some cases of hydrate, the use of slow penetration rates was adopted to manage the hydrate formation risk [7].

7.2.3   Hydrate Sediment Drilling Scenarios

There are several instances of hydrate drilling sediments for exploration of hydrate sediments and possible understanding of their production methods. Some practical field instances of using drilling mud for drilling hydrate sediments are discussed herein. Usually, when drilling gas hydrate sediments for exploration purposes, seawater is used as drilling mud since the focus is just to confirm the lithologies of the well. However, for methane production drilling purposes, in the Mallik Field (permafrost). Lecithin water-based mud was used because it has a strong ability to manage hydrate instability. The use of KCl polymer-based mud has been used in Nankai Field. In the same Nankai field, sepiolite drilling mud was also employed in some gas hydrate well drilling operations [8]. The use of sepiolite drilling mud is preferred to KCl/polymer mud [9]. This was confirmed via a riserless drilling of gas hydrate in the NGHP Expedition in India. The sepiolite mud used for drilling gas hydrate sediments is mostly combined with brine [10].

According to Grigg and Lynes [11], oil-based mud could manage hydrate risk better than water-based muds. The used of oil-based mud to drill a well in the Ignik Sikimu field for the testing of the swapping production method of CO_2–CH_4 (permafrost). The advantage of using oil-based drilling mud was based on the fact that they negligibly affect hydrate stability. Also, oil-based mud minimizes the risk of freezing in the drilling mud while drilling [10].

References

1. Englezos P (1993) Clathrates hydrates. Ind Eng Chem Res 32:1251–1274
2. Dzialowski A, Patel A, Nordbo MLLCK, Hydro N (2001) The development of kinetic inhibitors to suppress gas hydrates in extreme drilling conditions. In: Offshore mediterr. conf. exhib. Ravenna, Italy, pp 1–15
3. Barker JW, Gomez RK (1989) Formation of hydrates during deepwater drilling operations. J Pet Technol 41:297–301
4. Hale AH, Dewan AKR (1990) Inhibition of gas hydrates in deepwater drilling. SPE Drill Eng 5(109–115):18638
5. Ouar H, Cha SB, Wildeman TR, Sloan ED (1992) The formation of natural gas hydrates in water-based drilling fluids. Chem Eng Res Des 70:48–54
6. Harun AF, Krawietz TE, Erdogmus M, America BP (2007) Hydrate remediation in deepwater gulf of Mexico dry-tree wells : lessons learned. In: Offshore technol. conf. 2006. Houston, Texas, U.S.A, pp 1–4
7. Schofield TR, Judzis A, Yousif M (1997) Stabilization of in-situ hydrates enhances drilling performance and rig safety. In: SPE Annu. tech. conf. exhib., pp 43–50
8. Takahashi H, Co E, Yonezawa T, Corporation O (2001) OTC 13040 Exploration for natural hydrate in Nankai-trough wells offshore Japan. In: 2001 Offshore technol. conf. Houston, Texas, U.S.A, p 13040
9. Merey Ş (2019) Evaluation of drilling parameters in gas hydrate exploration wells. J Pet Sci Eng 172:855–877
10. Merey Ş (2016) Drilling of gas hydrate reservoirs. J Nat Gas Sci Eng 35:1167–1179
11. Grigg RB, Lynes GL (1992) Oil-based drilling mud as a gas-hydrates inhibitor. SPE Drill Eng 7:32–38

Printed in the United States
by Baker & Taylor Publisher Services